T0219476

A World Without Soil

A WORLD WITHOUT SOIL

THE PAST, PRESENT, AND PRECARIOUS FUTURE OF THE EARTH BENEATH OUR FEET

• • •

JO HANDELSMAN

WITH RESEARCH AND CREATIVE
CONTRIBUTIONS BY KAYLA COHEN

Yale
UNIVERSITY PRESS

New Haven and London

Published with assistance from the Alfred P. Sloan Foundation Program
in Public Understanding of Science and Technology.

Published with assistance from the Louis Stern Memorial Fund.

Yale University Press books may be purchased in quantity for
educational, business, or promotional use. For information, please e-mail
sales.press@yale.edu (U.S. office) or sales@yaleup.co.uk (U.K. office).

Set in Adobe Garamond type by Integrated Publishing Solutions,
Grand Rapids, Michigan.
Printed in the United States of America.

Library of Congress Control Number: 2021935605
ISBN 978-0-300-25640-6 (hardcover : alk. paper)

A catalogue record for this book is available from the British Library.

This paper meets the requirements of ANSI/NISO Z39.48-1992
(Permanence of Paper).

10 9 8 7 6 5 4 3 2 1

For John Nagy, and all the other farmers

Contents

Preface

I wrote this book because the plight of the world's soils is a silent crisis. Most people are unaware that the very ground beneath us is slipping away at an alarming pace. That thing we walk all over, call dirt, and expose to great indignities is at risk. As soil erodes, it destabilizes food production and environmental health. If the world takes action now, we can steward our soil into the future and continue to grow food for a burgeoning population while using soil as a burial ground for carbon, thereby reducing greenhouse gases and slowing climate change.

Human beings as a species have a habit of waiting too long to fix things. We ignore, debate, and doubt until we are imperiled and in crisis. But once we decide to act, we often bring ingenious solutions and a collaborative spirit to the job. The uplifting part of the soil crisis, as opposed to so many others we face, is that *we can fix it.* From experience we know how to practice agriculture in ways that will preserve soil for millennia rather than see it washed into rivers and swirled up into the atmosphere.

Soil is not just utilitarian. I also wrote this book to share my fascination with soil and the science underlying its genesis and power. I love soil's mystery and intoxicating properties. I'm mesmerized by its variety of smells, textures, and meanings. And the way life springs from it is like an annual miracle. But it's not a miracle—it's even better—*it is science.* The many ways that soil provides succor, from food to life-saving drugs, energize my study of it and foster my attachment to it.

There is still time to save this precious resource. Once you've read this book, I hope you will be inspired to take action to save our soil.

Prologue

Dear Mr. President,

I write to alert you to an emerging crisis that threatens soil across the United States and civilizations worldwide. Yes, we're talking about soil—we also call it *dirt, suelo* in Spanish, *Łeezh* in Navajo, *adama* in Hebrew, *talaj* in Hungarian, and *udongo* in Swahili.

The top layer of fertile soil, on which every form of life depends, is fast eroding. Soil is the product of millennia of physical forces acting on Earth's crust. Weathered geologic material mixes with chemicals released from dead and living plants, animals, and microorganisms. These are the basic materials of soil. Water percolates through, air fills empty pockets, plants penetrate, animals burrow, and microbes turn the cranks of nutrient cycles. Over thousands of years, soil is enriched and deepened by these processes to produce the fertile topsoil that we rely on for 95 percent of our food production.[1]

Soil's profound impact extends beyond agriculture. All organisms rely on soil for clean water—in fact, soil serves as the largest water filter on Earth. It is also the largest terrestrial repository for carbon, containing three times the amount in Earth's atmosphere and four times as much as in all plants combined, making it a powerful mitigative tool to address climate change.[2] As the most biologically diverse habitat on Earth, it contains microorganisms that are the source of drugs used in traditional and modern medicine. As a substance of complex physical properties, it is baked into bricks, roads, and pottery.

Worldwide, soil is under threat. Its erosion and degradation —already rapid—will likely accelerate with the projected rise in frequency of rainstorms generated by a warming climate. The United States and many other countries are eroding soil ten to one hundred times faster than it is produced. By some estimates, the United States will lose so much of its soil on sloped agricultural land that by the end of the twenty-first century, crop yields will be severely compromised. Some regions will soon be barren. Indeed, aerial photographs of Iowa farmland show frequent patches where rocky subsoil is already poking through the land surface.[3]

There is a long history of civilizations collapsing because of soil erosion. The number of Easter Islanders declined from fourteen thousand to two thousand after their soil eroded from steep mountainsides into the ocean, leaving the island without the capacity for agricultural production.[4] Similar examples of communities that have overtilled their soil, which then eroded, along with the ability to produce food, abound in China, Africa, and the United States. There is ample evidence to indicate that large tracts of U.S. cropland are following this trajectory. The current trend is not sustainable. If it continues, we will lose enough soil to create unprecedented challenges to food production.

The good news is that we have sufficient knowledge to diminish or even halt soil erosion with relatively little short-term cost and substantial long-term savings. No-till planting, use of cover crops, and interplanting crops with deep-rooted prairie plants compose the trio of proven agricultural practices that can prevent erosion and rebuild soil health. These methods of farming would enhance soil carbon storage, thereby reducing greenhouse gases. In the Paris climate talks of 2015, a proposal to increase soil carbon worldwide by 0.4 percent annually was advanced; though the goal is aspirational, if it were attained, sufficient carbon would be sequestered to compensate for projected increased carbon emissions and keep atmospheric carbon at current levels.[5]

There are several policies that your administration could implement to encourage farmers to adopt soil-protective practices and build soil carbon, from changing how crop insurance is calculated to reinstating accountability for farmers receiving incentives for soil conservation. The administration could galvanize consumers to participate in a movement toward a "soil safe" label for food and partner with farmers, environmental groups, agrichemical companies, food retailers, and indigenous peoples to develop criteria for certification.

I deliver to you a challenging problem, but one that can be remedied quickly. All we need is the will, and we dare not lack it or our civilization as we know it will be unsustainable.

Jo Handelsman
Associate Director
Office of Science and Technology Policy
2016

This is the memo I would like to have sent to President Obama. The rest of this book is my memo to all of you.

1

Dawning—An Invisible Crisis

How did I miss it? That was the question hounding my thoughts as I paced my White House office in 2015. There I was—a science adviser to President Barack Obama and thirty-five years a soil scientist—and I had somehow missed the fact that we were facing a crisis. Disappearing many times faster than it was being replaced, the soil of the United States could actually run out.

Let me back up a bit. Two years earlier, I had received a call from John Holdren, President Obama's top science adviser and director of the Office of Science and Technology Policy. He asked me to join him at the White House to run the office's Science Division, and initially I declined, reluctant to leave my bustling research laboratory at Yale University, where my undergraduate, graduate, and postdoctoral students probed microbial communities in insect guts, in soil, and on plants. But John's brilliance and the president's deep commitment to science won me over, and I agreed to serve. After the standard but grueling FBI background check, a crash course on how to answer senatorial questions, a surprisingly fun confirmation hearing in front of the Senate Committee on Commerce, Science, and Transportation, and a nine-month wait for a vote, I was finally confirmed by the full Senate and sworn in as the associate director for science.

The day of my swearing in, I moved into my majestic office in the Eisenhower Executive Office Building, where almost a thousand staffers were based, next door to the White House. I quickly found out that I had the run of the sixteen-acre White House complex. Since I rarely said good-night to the friendly Secret Service agents and passed through the security gates before dark, the lush beauty of those sixteen acres became my conduit to nature, replacing the field research and gardening that had grounded me throughout my adult life. The expansive lawns served as the setting for the president's formal military reviews with foreign leaders, his *we-need-to-discuss-a-serious-problem* walks with his chief of staff, and his impromptu sledding with the First Daughters during a snowstorm that shut down Washington. The grounds became my connection with land and history. As I hurried to and from meetings, I thought of the thousands of White House staffers who had walked these same paths—perhaps there were even molecules of past presidents drifting around me. I would smell the rich earth and compost as I passed the First Lady's vegetable gardens, which my staff vied for the privilege of weeding. I reveled in the colorful tidiness of the flower gardens, never spotting a sickly plant. And I would nod to the perfect rose bushes lining the White House colonnade that served as witnesses to my favorite treat (escorting the president from the Oval Office to events in the East Room).

As I adapted to my new position, the ornate walls of my office filled with images of satellites, telescopes, microbes, and the president—mementos of the projects that filled my days. My job was to serve the American people with science-based policy that would strengthen the research enterprise and improve the world. In practice, my division's responsibilities ranged from advising the president and the Office of Management and Budget on the annual science budget to overseeing the management of major scientific instruments such as telescopes and supercolliders. I was visited by hundreds of scientists who had visions of what should be done to strengthen science and its applications. I traveled abroad to discuss international initiatives that would support big scientific projects, leading several U.S. delegations to gatherings of science ministers from the G7, G20, and European Union. On one trip to Japan, I drove futuristic electric vehicles and held conversations with robots, but the truly thrilling (and terrifying) moment was when the minister chairing the meeting turned to me and intoned, "And what does the United States think?" It was up to *me* to convey the White House's position to the other countries.

Working in the Obama administration was exciting. I learned about searches for the tiniest particles of matter and for new galaxies in distant parts of the universe, the quest for universal vaccines against viral diseases, and the mission to save the Venus flytrap. The rewards were fantastic. I saw the president announce, in the 2015 State of the Union address, his intent to implement the precision medicine initiative that I had developed with colleagues in OSTP and at several federal agencies. Thanks to the extraordinary White House legislative staff and a deep commitment in Congress to human health, the initiative was ultimately funded in a bill that passed the Senate ninety-two to eight—now that was bipartisanship at work!

There were also moments filled with tragedy and despair, such as when we learned of the Ebola and Zika epidemics. But there was always hope: we were, after all, the government of the United States of America, and we should be able to help alleviate pain and avert death, right? When the Ebola epidemic started in West Africa, the president made it clear that he wanted the outbreak extinguished. Fast. I spent many months attending meetings in the Situation Room; watching in admiration as the U.S. Army constructed treatment units across Liberia in three weeks; holding discussions with staff at twenty-six federal agencies; witnessing the monumental efforts of President Obama's legislative staff to obtain financial backing from Congress; making some missteps, having arguments, being bested in power struggles; and meeting the brave health care workers who saved thousands of lives. Then, one day, Liberia was declared Ebola-free. I was so proud to be a tiny part of the U.S. government that made that happen.

When I wasn't responding to crises, requests from the president, and the demands of science diplomacy, I worked on my own science agenda. A chief objective was to increase the profitability and sustainability of the U.S. food supply. My career in agricultural sciences had shown me that climate change and economic shifts were creating new, sometimes unbearable, pressures on American farmers. Two needs of the agricultural enterprise were particularly acute. The first was the need for more plant breeders, a dying breed of scientists that we depend on to improve crop production through genetic selection. The second was the need to improve the quality of soil used in crop production. The first challenge involved expanding training, and the second, I thought, would require a research initiative.

As I brushed up on modern soil science, however, I discovered a disconnect between research and practice. Much of what soil scientists had learned in recent decades about improving soil quality was not being implemented on two-thirds of U.S. fields, leading to an insidious problem: our nation's soil was disappearing. Whisked away by wind and water, flotillas of soil, though largely invisible, were moving across the United States. Soil erosion was especially rampant in the midwestern states of Minnesota, Iowa, Kansas, Arkansas, Missouri, and Illinois, where each year tons of topsoil entered the Mississippi River and flowed to the Gulf of Mexico.[1]

This came as a surprise. I had studied soil science in the 1970s and followed trends in soil erosion until 1985, when Congress passed the National Food Security Act, which launched a vast soil conservation effort managed by the U.S. Department of Agriculture's Natural Resources Conservation Service (NRCS). I knew that over the ensuing few years, great strides had been made in curbing soil erosion due to the conservation service's efforts. After that, I focused on my research in soil microbiology and lost track of national trends, assuming that the problem of soil erosion had been conquered.

What had simply passed me by was that after 1985, Congress weakened the language of the legislation repeatedly until the NRCS had little power to hold farmers accountable for soil protection. By 1992, progress to reduce soil erosion had slowed considerably. Today, the United States is losing soil ten to one hundred times faster than it is produced. The international situation is worse.

I was appalled by what I discovered in published research on soil erosion. And what I learned in conversations with renowned soil scientists not only confirmed my fears but warned me that the reality was even worse than many studies suggested. Using USDA data, my staff and I made projections that showed how dire the situation was: large tracts of midwestern farmland could be devoid of topsoil within the century. When we accounted for the increases in heavy precipitation predicted by climate change models, the rate of soil erosion accelerated considerably. Iowa State University agronomist Rick Cruse shared with us an aerial photograph showing many spots where the topsoil was gone, baring the subsoil to the elements.

For months, I rechecked my calculations, verified the data, read further scientific studies, and consulted more experts. Only when I was steeped in the science did I accept that the word "dire" was not hyperbole. This soil erosion trend could reduce America's abundant food production in a matter of decades, and similar rapid soil loss in India, China, and Africa could spark an international food crisis.

As I paced my office, dismayed by my former ignorance, I considered what to do. I had to tell the president. How could I be a responsible adviser and keep this catastrophe from him? Saving America's soil could become a legacy of the Obama administration.

Any process to inform the president begins with a memo. Most of my prior memos had reached him, and some had been followed by a meeting in the Oval Office to consider next steps, so I felt optimistic that this memo would be received by President Obama, discussed, and result in a presidential initiative on soil preservation. But after waiting over a year, I was finally told that it was simply too late in the presidential term to start another major initiative. It remains my biggest regret that I was never able to get that memo to him, but I can still take action by writing this book to inform people around the world about the crisis.

The soil crisis is real and coming fast and will ultimately affect every living thing on Earth. Erosion is progressing at different rates across the globe, but like a spreading virus, soil erosion doesn't discriminate, nor does it affect only those who experience it up close. The loss of soil will influence our access to food and drugs and will change Earth's climate. It will eradicate some habitats and expand others, and it will modify the distribution of species, triggering extinction of some and surges of others. But soil erosion can be stopped, and quickly. Humans should be motivated by the hope and empowerment that is intrinsic to a crisis we can reverse. Let's learn more so that we can act with purpose and unity.

2

Earth's Dark Matter

It's no coincidence that we named planet Earth with a synonym for "soil." There's more to our world than its surface, but that thin brown skin of minerals and biomolecules is a defining feature, unique within the known universe, and worthy of our planet's title. If it's worth the title, then soil's story is worth telling from the beginning.

Earth formed long before there was any earth, or soil, present. The Big Bang theory is the best approximation for how the universe began and, by extension, how Earth formed. The theory holds that in the beginning all the matter and energy of the Universe was packed into an infinitesimally tiny point that exploded 13.7 billion years ago in an event we now call the Big Bang. In its first moments the universe was small, dense, and so hot that it was impossible for particles to form. As the universe's temperature dropped, first to form were quarks and electrons, followed by the simplest of the 118 elements: hydrogen atoms, comprising a proton and an electron. When the temperature fell below 1 billion degrees, hydrogen nuclei fused to create helium. Further cooling led to giant clouds of hydrogen and helium gas called nebulae, and in their misty depths, the first galaxies erupted with stars.

Planetary systems mature over millions of years as gravitational and magnetic fields collapse nebulae into protostars, which then accumulate discs of cosmic debris that spin and smash together to form rocks, protoplanets, and ultimately planets. In our solar system, one of the eight planets became Earth around 4.55 billion years ago. Minerals dating back 4.4 billion years ago are found within granitic rocks, which require water to form, and so geologists infer the presence of rock and water as early as 150 million years into Earth's existence.[1] Water and rock are two critical ingredients of soil. The third ingredient is life.

The earliest fossil and rock records place the origin of life on Earth somewhere between 3.95 and 3.48 billion years ago, but it might be even earlier.[2] Life can be defined as a self-sustaining chemical system capable of change. The first chemical systems resembling life on Earth were probably self-replicating molecules of RNA (single strands of nucleotides closely related to DNA), followed by simple single-celled organisms that were the ancestors of Bacteria and Archaea—two of the three domains of life. These early unicellular forms of life likely originated in the boiling hydrothermal vents on the seafloor. The process by which they formed remains a mystery.

As soon as life emerged it began changing by processes such as evolution by natural selection. Populations contain individuals with slight genetic differences. If a variant gene confers a fitness advantage enhancing an individual's likelihood of survival and reproduction, then that variant will become prevalent as the less fit variants decline.

The process of change was reciprocal—the environment acted as a selection pressure on life, and life-forms changed their environment by altering Earth's chemistry and atmosphere. Both processes have continued for over 3 billion years. By far the most dramatic example of life-forms transforming the Earth was the production of oxygen by cyanobacteria that led to the Great Oxygenation Event of 2.45 billion years ago.[3] These photosynthetic bacteria used the Sun's energy to extract carbon from atmospheric carbon dioxide, releasing oxygen as a waste product. Once the oceans and the minerals in them became saturated with oxygen, it accumulated in the air. As cyanobacteria thrived, more and more oxygen was released, bringing the atmospheric concentration to levels sufficient to support respiring forms of life. In a stark example of evolution by natural selection, many anaerobic species that could not tolerate oxygen became extinct, setting the stage for life-forms capable of metabolizing oxygen. Today, almost all complex organisms demand oxygen, illustrating just how influential microbes have been, affecting the metabolism of millions of species over the course of evolution.

Cyanobacteria were responsible for another remarkable change, this time in the stratosphere. As oxygen accumulated in the atmosphere, it diffused into the stratosphere, where the concentration became sufficient to drive the formation of the ozone layer. Ozone is formed when ultraviolet light from the Sun splits O_2 molecules into single atoms of oxygen. If one of these atoms collides with an O_2 molecule, ozone—or O_3—can form. When enough of them formed, they created a filter that absorbs enough of the Sun's ultraviolet rays to protect life on Earth from getting zapped by deadly radiation. These events set the stage for plant, animal, and microbial life to diversify and migrate onto land, laying the foundation for soil.

How do we know what happened billions of years ago? One of the fascinating parts of science is how we learned what we now know. I started out practicing science in an experimental paradigm—we manipulate variables, conduct experiments, and look for outcomes. If we change the temperature, what will the bacteria do? If we remove a gene, how will the plant function? If we administer this drug, will the patient recover? All experiments contain controls that provide a baseline for comparison so that we can measure the impact of the manipulated variable. And all experiments are replicated: we don't use results from one bacterium, plant, or person but rather gather many results to determine how much the effect of interest varies among individuals or even for the same individual on different days. Replicated experiments with controls were part of my definition of science; events needed to be controlled and repeatable to be meaningful. So when I first started learning about the formation of the Earth and its soil, I was flummoxed by the challenge of having only one Earth. One Earth means no replications. And all of its formation was, well, in the past, which meant no experiments to see whether, under changed conditions, Earth would form the same way. This meant that scientists studying events long past couldn't replicate experiments or even do experiments of the type I do in my microbiology lab. Planetary science involves an entirely different way of knowing.

I have come to admire the thinking and inference involved in constructing a picture of early Earth and the forces that created it. Take the example of figuring out when the first two ingredients of soil—rocks and water—first came to be on Earth. Geologist John Valley at the University of Wisconsin–Madison revolutionized his field and rewrote the textbooks when he discovered evidence that the Earth cooled dramatically 4.4 billion years ago, scaling back estimates of the formation of the continental crust and the appearance of oceans by over 400 million years.

Valley was four years old when he first approached the granite quarries of southern Maine with a small hammer and look of determination that made his father laugh. Every summer they would travel from Boston to the Maine quarries where little John could mine for crystals and, if he was lucky, gemstones. Decades later, he still loves rocks. As a geologist Valley says that he's never worked a day in his life, which is code for saying he'll never retire.

John Valley studies zircon, a mineral he uses as Earth's timekeeper. Zircons are made of zirconium silicate (zirconium, silicon, and oxygen). The three elements and the crystallization process are common, so zircons are abundant in Earth's crust. Once zircons form, they persist. Under extreme heat and pressure, they persist. When their host rock is crushed and melts at high temperatures, zircon chemistry remains intact. But it is not zirconium silicate that holds tales of Earth's history; rather, the evidence derives from the molecule's propensity to capture *other* elements. This nifty property enables scientists to date the mineral's formation. As they form, zircons can trap radioactive uranium atoms. Over time the uranium atoms decay into an element with a larger atomic size—lead. Because lead atoms are too large to be trapped during the initial formation of zircon, they can only appear inside zircon crystals if they are derived from decaying uranium. Radioactive decay occurs at a known rate, and so it is a simple calculation to convert the uranium-to-lead ratio to an age, with pure uranium content indicating the youngest crystals and high lead amounts indicating the oldest. Using this method, Valley and his colleagues dated some very old zircons and found that they had formed 4.4 billion years ago.[4]

That was rock. The second component of soil is water. When did the first oceans (the earliest bodies of water on Earth) appear? To infer the conditions surrounding the formation of the ancient zircons, Valley's group relied on isotopes, variants of an element with different numbers of neutrons in the nuclei. In this case, Valley was interested in the isotope ^{18}O, which has two extra neutrons in its nucleus than the majority of oxygen, ^{16}O. Both ^{18}O and ^{16}O are stable over time, meaning they do not decay into other isotopes over time. The ratio of ^{18}O to ^{16}O in zircon is determined by the temperature at which the crystal formed, and for Valley's zircons, more ^{18}O indicated lower temperatures. The isotope ratio thereby acts as a thermometer that permanently locks in the temperature reading. Valley found that the ratio of ^{18}O to ^{16}O was very high in the 4.0- to 4.4-billion-year-old zircons, indicating that they formed from rocks altered at a relatively low temperature (probably in soil) and then buried to melt at high temperatures to form the zircons. In fact, the weathering was at temperatures low enough for water to be liquid, thereby placing the emergence of oceans over 400 million years earlier than generally thought.[5] Isn't that an amazing set of deductions derived from grains of zircon sand in a handful of rocks?

At a mere 150 million years old, early Earth had established two of three ingredients for soil: rocks and water. Life was the last ingredient.

Since 2005, John Valley has applied his powerful tools and keen insight to clocking life's origin using microfossils. Until 1992, scientists could only guess that life was around two billion years old because there was no method to trace it directly.[6] Then paleobiologist J. William Schopf discovered what he proposed was the oldest evidence for life on Earth embedded in ancient rocks in Western Australia. But many in the field were skeptical of his claim because it was based solely on visual analysis of microfossils. How could something as important as the origin of life be decided by details of shape, color, and structure? Morphological observations were important, but not enough to rule out hypotheses that these specks were nonbiological minerals.

For ten years, Valley tried to persuade Schopf to share the precious microfossils he had found in a 3.465-billion-year-old rock, but the key samples were curated in a museum in London. When he finally obtained them, Valley conducted a formidable chemical analysis that demonstrated that microfossils with different morphology also differed in their isotope content, indicating that they were different species. The collaboration established that these microfossils were, indeed, remnants of organisms that had lived 3.465 billion years ago, most likely in mud or underwater.[7]

And so we know that within one billion years of its formation, Earth was already teeming with diverse microbial communities, each species sporting its own metabolic secret for thriving in a competitive world. By its billionth revolution around the Sun, Earth had developed all the components needed for soil formation (fig. 1).

Soil forms in a millennial process driven by weather and microbial activity. It starts with rocks. The geologic base, known to soil scientists as parent rock, is the first defining feature of the soil that will be born from it. Rocks are weathered by thermal, mechanical, and chemical processes. Heat causes rocks to expand and fracture. Water penetrates rock crevices, and when temperatures fall below freezing, its liquid structure expands into solid ice, cracking rocks further. Later, plant roots work their way into crevices with sufficient force to split rocks, exposing more of their surfaces to the environment. Exposure can change rocks by diverse chemical means—some rocks, such as limestone, dissolve over time in water. Others are chemically altered by oxygen and elements in nearby rocks. The loss or

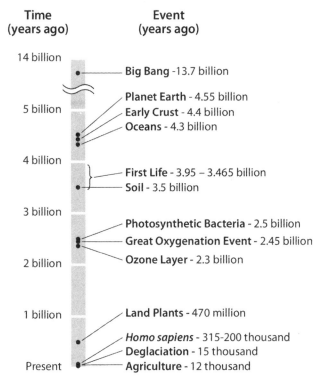

Time | **Event**
(years ago) | (years ago)

14 billion

Big Bang -13.7 billion

Planet Earth - 4.55 billion
5 billion
Early Crust - 4.4 billion
Oceans - 4.3 billion

4 billion

First Life - 3.95 – 3.465 billion
Soil - 3.5 billion

3 billion

Photosynthetic Bacteria - 2.5 billion
Great Oxygenation Event - 2.45 billion
2 billion
Ozone Layer - 2.3 billion

1 billion
Land Plants - 470 million

Homo sapiens - 315-200 thousand
Deglaciation - 15 thousand
Present
Agriculture - 12 thousand

Figure 1. A timeline of key events contributing to the formation and use of soil on Earth. Illustration by Bill Nelson.

gain of electrons changes the reactivity of elements, particularly metals, on rock surfaces. These influences break rocks into smaller pieces, change their colors and textures, and alter their chemical properties.

Most of these processes are regulated by weather events. Weather shapes soil formation by exposing parent rock materials to water and temperature fluctuations. Over time, rocks crumble into something resembling inorganic soil. The aggregate of weather events over a fifty- or hundred-year period is what we call climate. Earth's climate, which swings between greenhouse and icehouse states, also drives the behavior of glaciers, great juggernauts of ice that carve valleys and pulverize rocks as they make inexorable advances and retreats across continents. If it takes a puff of wind to displace soil, it takes a glacier to lift boulders and grind them along Earth's surface for millions of years, leaving behind a trail of pebbles.

Rocks differ in their rates of change. In abundant water, limestone dissolves rapidly, geologically speaking, whereas quartz, one of the hardest and least porous of geologic materials, is usually the last rock remaining. It follows, then, that soils made from different parent rocks will vary in their rates of formation and erosion.

Every handful of soil tells an anthology of stories, each belonging to a constituent mineral. Take silicon dioxide, the most abundant substance on Earth's surface. Its remarkable properties have made it both a bedrock of the Earth's crust (sandstone) and a workhorse of modern technology. Silicon dioxide is the chemical that comprises sand found on beaches, in deserts, and as a critical component of soil. Around 6500 BCE, early Bedouins found that when they mixed sand with lime and water, it produced a moldable blend that hardened to a sturdy building material similar to modern concrete. The Romans perfected the method, creating virtually indestructible buildings such as the Pantheon and Coliseum, both still standing today.[8]

Between 4000 and 3500 BCE, people in Egypt and Mesopotamia discovered that sand melts at high temperatures and, when cooled, hardens into a beautiful, useful substance we call glass.[9] For thousands of years, people have been fascinated with the properties of glass, adding minerals to alter its properties, making it clear for windows and colorful for art. In the past two centuries people found that sand strengthens asphalt roads and began laying down vast expanses, threatening to exhaust the world's sand supply. Although that sounds preposterous, given the abundance of deserts on Earth, only sand grains with angular shapes are useful as building materials, and desert sand has been worn smooth by millennia of exposure to wind. Most famously, silicon dioxide has become a coveted mineral for production of computer chips, which explains the naming of the computer industry near San Francisco, *Silicon* Valley.

Given all of its impressive properties, it is no wonder that silicon dioxide forms a critical part of many soils. Sand grains are the largest mineral particles in soil, imparting porosity that enables air and water to move freely. The smallest particles in soil are clay, which can be as tiny as a bacterium and are most often composed of silicate minerals with water trapped inside. These silicates mix with metals and are altered by reactions with water and oxygen. The metallic components, often aluminum and iron, create the rainbow of colors associated with clays—from the red bluffs of Martha's Vineyard in the United States to the tan cliffs of Omarama, New Zealand.

Midsize particles of soil are called silt. Each silt particle contains a mixture of quartz and feldspar, both products of cooling magma, the molten material under the Earth's crust. Silt also contains aluminum-silicon-based compounds bound with potassium, sodium, or calcium. These minerals are embedded in rocks that are gradually weathered by water and ice. Moving water dislodges rocks, and during their aquatic journey, fragments chip off, scrape against riverbeds, and grind against each other until they are worn down to silt particles.

Within this mineral infrastructure is a metropolis of microorganisms cooperating and competing for resources in unimaginably close quarters. There are an estimated 3×10^{29} bacteria living in soils worldwide, which equates to a few billion in a teaspoon of soil, and the weight of about five cows in a hectare.[10] Just as cyanobacteria demonstrated that microbes can drastically affect Earth's atmosphere, so soil chemistry proves the profound influence of underground microbes. Bacteria related to those that oxygenated Earth's atmosphere continue to fix carbon today, removing carbon dioxide from the atmosphere, using it themselves, and making it available to other soil dwellers.

Microorganisms serve as stalwart partners throughout mineral weathering events, accelerating spontaneous chemical reactions and adding their own chemistry to the mix. For example, certain bacteria produce acid that dissolves minerals into their nutrient components; others produce chelators, molecules that remove metals from the geologic material. These chemical processes can result in visibly distinct layers in meter-deep soil profiles. Of particular importance to soil and plant health are bacteria that convert dinitrogen gas, a biologically inert molecule from the atmosphere, into ammonia, a form of nitrogen accessible to plants. These nitrogen-fixing bacteria provided the main source of nitrogen in ancient ecosystems that enabled plants to establish on land.

Today land plants are the most abundant form of life on Earth, representing 450 gigatonnes (billion metric tons) of carbon. It's hard to imagine Earth without plants. They are the stewards of the soil, helping to create, shape, and nurture it. But taking the evolutionary leap from aquatic to terrestrial environments wasn't easy. As those hardy survivors began to grow in pulverized rocks, an ecosystem developed with contributions from many members. Bacteria provided nitrogen and other nutrients in forms usable to plants, and plants fixed carbon and fed it to microscopic and macroscopic organisms living on the rocky substrate. Plant roots pump into the soil between 20 percent and 40 percent of the carbon fixed through photosynthesis, making them the primary providers of carbon to the soil, which in turn makes soil the most biodiverse habitat on Earth.[11]

Plants changed soil at least 470 million years ago when they first migrated from sea to land, and change continues every time pristine rock encounters a plant. I was fortunate to see this process in action in 1985 when I attended a scientific meeting on nitrogen-fixing bacteria in Corvallis, Oregon. The meeting organizers arranged a trip to nearby Mount St. Helens, the site of a recent volcanic eruption. On May 18, 1980, pressure from deep inside the Earth forced magma to the surface, producing a bulge on the side of the mountain, which then blew with a force equivalent to sixteen hundred atomic bombs the size of the one dropped at Hiroshima. I had watched footage of the Mount St. Helens eruption on the news, which reported eerie, nightlike skies at midday that extended for hundreds of kilometers around the volcano after it spewed 520 million tonnes of ash more than 24,000 meters into the stratosphere, blotting out the sun. Poisonous volcanic gases released into the atmosphere killed fifty-seven people and thousands of animals.[12] Ash shut down air traffic in the Northwest, clogged waterways and farm machinery, and threatened the simple act of breathing. Over the following two weeks the ash surrounded the planet, creating stunning sunsets around the globe— a fiery reminder of the explosive energy prowling beneath the Earth's surface.

When I visited the volcano in 1985, the lava had hardened into a vast moonscape, extending into a seemingly bottomless crater. As our group skirted the crater's rim, our guide pointed to a tiny, lone plant growing in the lava rock. The nerdy bacteriologists gathered to peer, not at the barren majesty of the crater on whose edge we balanced, but at the plant, realizing that we were seeing our very own science in action—the little pioneer was a legume, a resolute plant that hosts nitrogen-fixing bacteria in nodules on its roots. Without nitrogen-fixing bacteria, plants cannot grow in lava rock for lack of nitrogen. As this plant grew, the lava rock would weather and crumble, more bacteria would thrive, and diverse plants and microscopic animals would join them. When these plants, animals, and bacteria died, other microorganisms would decompose their remains, and slowly, iteratively, an ecosystem called soil would form.

We were mostly silent for the rest of the hike, awed by the resilience of life and by the profundity of endless cycles of eruption and death giving way to new communities. As we turned to leave, a ground squirrel scampered around the crater's rim.

Animals of many sizes—from microscopic worms to burrowing badgers—play a vital role in the formation of soil. Below the Earth's surface, myriad soil habitats shelter about 25 percent of the world's described species, providing crucial reservoirs of biodiversity and, thus, ecosystem functions. A single square meter of soil can contain more than 150 individual earthworms with a collective biomass that can exceed 1,500 kilograms per hectare, roughly equal to the weight of two cows.[13] Tiny insects also settle underground, if not permanently, then as tourists. Ladybird beetles are transient residents, checking into the soil hotel only to hibernate. Certain moth larvae are residents only by day and leave to feed on plants by moonrise. And then there are the larger residents of mammalian descent that burrow in soil for food and shelter. All of soil's residents contribute to its structure by cementing particles or forging pathways for movement of air and water.

Since its genesis, soil has been the gift that the Earth keeps on giving. The minerals create a mixed substrate that is acted upon by water, microbes, plants, and animals. Water transports particles and organisms downward. As plants develop a foothold, they slough off polysaccharides (long, sticky strings of sugars), protein, and DNA that together form a thick mucus that binds soil particles into aggregates or clods and makes a wholesome broth for microbial residents. Microorganisms transform minerals to different chemical states, making nutrients such as nitrogen, phosphorus, and magnesium sometimes more and sometimes less accessible to plants. As plants, animals, and microbes die, their remains are decomposed by fungi and bacteria that chew large polymers into small molecules of food, providing them energy to synthesize chemicals needed by their own cells while building soil structure and richness.

As the most complex habitat on Earth, soil possesses mystery and power unequaled by any other resource. No plant or animal, oil or coal deposit, waterfall or towering mountain contains the vital force or intricacy of soil. And few features of the world are as hidden from view, as humble in nature, or as taken for granted by those who depend on it for survival. And few are as challenging to study. Soil is the dark matter of our planet.

3

Earth Works

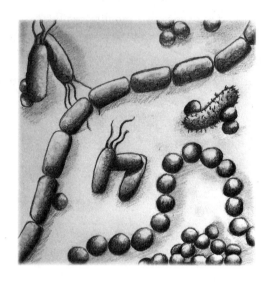

Walk outdoors on a spring day just after a rain, take a deep breath, and you have a good chance of smelling soil at work. That rich smell emanating from the soil is geosmin, which literally means "earth odor" in Greek. Geosmin is a chemical released by soil-dwelling bacteria, its richness giving voice to the industrious microorganisms toiling in the earth to create our food, water, fuel, building materials, and medicines (fig. 2).

Soil offers an eclectic repertoire of ecosystem services because it is a site of many biological, chemical, and physical processes. From a biological standpoint, soil equips plants and microbes to thrive, enabling the production of food, fuel, feed, fiber, and medicine. From

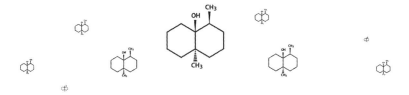

Figure 2. Structure of geosmin. Illustration by Bill Nelson.

a chemical standpoint, soil is the filter that snatches and discharges both beneficial and toxic chemicals as they pass through it. And from a physical standpoint, soil provides the structural support for plants while modulating downward water flow. Soil accomplishes countless tasks for Earth's inhabitants; perhaps that's why it's been so revered throughout human history.

Most polytheistic religions have a deity representing soil. In Greek mythology, Demeter, goddess of the Earth, promised abundant harvests if people managed their soil properly. In the Hindu pantheon, Earth is personified as the goddess Bhudevi. In Judeo-Christian theology, soil is not elevated to godly status, but the relationship between soil and human life is fundamental. The Old Testament states that Adam, the first man, was created from the dust of the Earth. The name Adam connotes the Hebrew word for soil, *adama.* Similarly, the Koran tells of Allah molding people from clay and mentions the word "soil" 287 times to describe sediment, land, and the universe.[1]

In 1912, Curley, a Crow Indian, explained why he was refusing to sell more land to the government. He illuminated the ancient bond of his tribe with soil thus: "The soil you see is not ordinary soil—it is the dust of the blood, the flesh and the bones of our ancestors. . . . The land as it is, is my blood and my dead; it is consecrated; and I do not want to give up any portion of it."[2] Curley's struggle to communicate with the U.S. government about soil is typical of the chasm of understanding between many indigenous peoples and industrialized nations, few of which have retained reverence for soil. Industrialization and centralization of food production has made it easy for people to live removed from the source of their food and to look on soil as "dirt," a word derived from the Old Norse *drit,* meaning excrement. Dirt? A word that means filth, corruption, obscenity, gossip, worthlessness? This is how some people refer to the treasure of the Earth's crust, the life-giving, labyrinthine habitat that provides us sustenance, polishes our water supply, fortifies us against disease, and stores vast amounts of carbon. Soil.

Just as the Great Oxygenation Event changed the Earth's face forever, so, too, did the transition of *Homo sapiens* from hunter-gatherers to agriculturalists. Before this shift occurred approximately twelve thousand years ago, soils had formed by natural processes for millions of years.[3] Whatever grew on a patch of land was native to that area or else had been dispersed by wind, water, or animals. Over the millennia, soil hosted much of Earth's history: it was the birthplace of countless plant species and the abiding graveyard of all those once-impressive terrestrial animals now extinct.

Between 5 and 7 million years ago, early hominins diversified across Africa. When the *Homo* genus left Africa for Eurasia 2 million years ago, they distinguished themselves from other hominin taxa. Between 200,000 and 315,000 years ago, *Homo sapiens* emerged. To this day, it is unclear why our species was spared the mass extinction that wiped out eight other *Homo* species. We know that the invention of stone tools long preceded our species, but perhaps it was our dexterous hand morphology that gave us an edge in designing and using technology and, consequently, an evolutionary advantage. Whether or not the propensity to rotate our wrists and carry more weight helped humans survive, it certainly helped us perform agricultural activities thousands of years later, which corresponds with a dramatic population increase from eight million people 10,000 years ago to almost eight billion today.[4]

Agriculture is a specialized activity in which humans and a few other animals, such as ants, manage and exploit organisms that serve as food. Whereas some archaeologists argue that the gathering of wild grains some twenty-three thousand years ago constitutes proto-agriculture, it's generally accepted that farming began eleven thousand years later, in the warmth that followed the last glaciation.[5] As the climate shifted, diverse plants spread, and *Homo sapiens* discovered the practice of collecting seeds from wild plants to cultivate as crops for the following season. Early farmers enhanced production by removing plants that competed with their crop. They also influenced plant genetics by selecting for favorable traits, such as larger seeds.

With agriculture came settlements, followed by villages and cities. People began to live in clusters, which eventually enabled diversification of occupations, the arts, and the emergence of a leisure class. Agriculture also facilitated the spread of infectious diseases in people, livestock, and crops—many individuals of the same species living near one another makes it easy for pathogens to hop from host to host. Disease, compounded by the vagaries of weather, led to fluctuations in food abundance, resulting in times of lavish plenty followed by scarcity. Agriculture nourished expanding populations as well as economies, creating incentives to exploit soil. Underlying the collapse of several ancient (and not so ancient) civilizations is degraded soil, shredded by plows, depleted by cultivation, and left vulnerable to erosion by wind and water. Agriculture led to a paradox at the heart of today's soil erosion crisis: farming can heighten people's awareness of the value of soil while fostering its abuse.

Today, soil supports production of 95 percent of the human food supply.[6] At the base of our food chain are plants that use energy from sunlight to drive photosynthesis, the process of converting atmospheric carbon dioxide into sugars. Soil provides plants with water, sulfur, potassium, magnesium, calcium, iron, phosphorus, and other nutrients necessary for photosynthesis and growth. As nutrients pass along the food chain, organisms inherit the elements needed to build biological molecules.

As I observed at the Mount St. Helens crater, leguminous plants are often the first to colonize barren rock. This colonization occurs because most of the nitrogen in the environment is inaccessible to plants and animals, and the nitrogen-fixing bacteria associated with legumes can access it. Although nitrogen makes up 78 percent of the air, very few organisms can use it in its gaseous form. Yet all organisms require nitrogen to build the molecules of life. There are only three processes that convert inert nitrogen gas to a form plants can use. The first is lightning strikes, which are sky-to-ground bolts of electric energy strong enough to kill thousands of people each year worldwide.[7] This provides a sense of the magnitude of energy needed to snap the nitrogen atoms apart. But lightning is responsible for only a small proportion of the nitrogen in biological systems. Certain bacteria can directly access the atoms in dinitrogen gas—two atoms of nitrogen joined by a triple covalent bond, one of the strongest bonds in nature.

$$N \equiv N + 3H_2 \rightarrow 2NH_3$$

For thousands of years, farmers relied on nitrogen-fixing bacteria to free nitrogen atoms and build ammonia (NH_3) and other nitrogenous molecules. Especially useful were the rhizobia, bacteria that fix nitrogen for leguminous plants such as peas, alfalfa, and soybeans.

These plants, because of their nitrogen-fixing symbionts, contribute to soil health. For centuries, farmers have been plowing residues of legume crops into soil, passing on nitrogen to subsequent nonleguminous crops such as rice, wheat, corn, and potatoes. Historical records indicate that even the ancient Romans transferred soil from old plots of peas to establish new pea crops in fresh soil, unknowingly transferring rhizobia.[8]

Until the twentieth century, lightning and nitrogen-fixing bacteria were the only nitrogen sources for life on Earth. That's worth a moment of thought. Every nitrogen atom in every organism on the planet—every petunia, giant sequoia, dinosaur, mosquito, cow, and human—had been freed from a dinitrogen molecule by nitrogen-fixing bacteria or, rarely, by lightning. Shouldn't soil bacteria get some respect for that work?

In the early twentieth century the relation between nitrogen and agriculture was reconfigured when German scientist Fritz Haber developed a method for converting dinitrogen gas to ammonia. In a search for ways to make explosives and nitrogen mustards for warfare, he discovered that nitrogen's triple bond would surrender under high temperature and pressure. The discovery became commercially viable when another German scientist, Carl Bosch, developed a catalyst that increased the efficiency of ammonia production to industrial standards. Together, their work constitutes the Haber-Bosch process, still used today to produce nitrogen fertilizers. Haber and Bosch each received the Nobel Prize in chemistry for their work because of its far-reaching impact on human life. How ironic that research initiated for the purpose of killing people should transform food production.

With the Haber-Bosch process, nitrogen fertilizers became commonplace in agriculture and boosted crop yields by at least 30–50 percent.[9] These fertilizers underpinned the Green Revolution of the

1960s, which bred high-yielding crops for low-yielding agricultural systems. Plants were bred under high-nitrogen conditions, enabling them to achieve high yields. Unfortunately, this achievement also made the future of agriculture dependent on nitrogen fertilizers and, by extension, fossil fuels, which provide the energy to split the triple bond in nitrogen gas. It's striking that nitrogen-fixing bacteria have no need of the power of fossil fuels to split the renowned triple bond; they carry out the reaction at standard temperatures and pressure, whereas the Haber-Bosch process requires temperatures above 200°C and pressures hundreds of times greater than ambient. Maybe soil bacteria should get some respect for that work, too?

It took me a long time to find out about these handy nitrogen-fixers. Before I studied them, I scrutinized all kinds of tiny creatures. I suppose it all started in my seventh-grade science class, where I first used a microscope. I'll never forget staring, mesmerized, at paramecia, adorable single-celled organisms, as they gobbled up crystals of carmine red, sweeping the crimson bits into their mouths with cilia waving in unison. I was late to my next class, but I didn't mind. I never wanted to part with my new friend, the microscope. I felt as though Nature had taken me into her confidence—that I'd been lucky enough to have a very big secret whispered to me in that science class.

Knowing that I had to learn more of these microscopic secrets, I saved up my babysitting money until I had enough to buy an old Leitz microscope that had somehow migrated from a 1930s German hospital to a stuffy warehouse in New York and finally into the hands of a twelve-year-old girl eager to see tiny things. For the next four years, I spent most of my free time gazing through the Leitz microscope at everything from pond water to crystals of vitamin C. I took delight in my old microscope's classic brass knobs and scratched black body as well as the world it allowed me to enter. When I started my own lab several years later and had the task of choosing a new, modern microscope for my research group, of course I chose one made by Leitz.

As a college student, I studied plants but had another one of those heart-stopping moments in a soil science class when we learned about nitrogen-fixing bacteria. As the professor described how rhizobia infect the roots of leguminous plants, inducing the formation of little organs called nodules, then fixing and feeding nitrogen to the host plant, I once again felt as if I'd seen the unseeable. Here was a view normally obscured by both the size of the subjects and that opaque, enigmatic habitat, soil.

Despite being captivated by rhizobia, I entered graduate school intending to study plants, but one research lab captured my imagination because of its creative and inspiring leader, Winston Brill, and his research on nitrogen-fixing bacteria. I settled into microbiology. That time, I finally received the message from the bacteria loud and clear: "Study *us*," they whispered, beckoning me away from plants. And I have never since turned my attention away from the microbes. Plants became accidental tourists in my grand tour of the world of bacteria, where I probe the secrets of the tiniest of organisms that live out of view, in soil or on roots. I credit the paramecia and nitrogen-fixing bacteria with getting me hooked on a lifetime of microbial curiosities.

Soil is a bustling marketplace where all the consumers, living and inanimate, exchange nutrients. It is in this marketplace that much of the global carbon and nitrogen economies are managed. As is so often true in microbial processes, economies of global impact are commanded at minute spatial scales. Carbon and nitrogen are managed in cycles—elements are repeatedly transformed into different molecules until eventually the element is returned to its original state. It's strange to think that Earth has been recycling its own atoms that amassed 4.5 billion years ago, losing only a fraction of its original stock to outer space or radioactive decay. Much of the Earth's goods will eventually pass through the soil marketplace to be disassembled piece by ancient piece.

Soil's intricate role in the carbon cycle is of particular interest today because of its effects on climate and agriculture. The process of carbon capture starts with plants. They use the carbon they fix through photosynthesis to produce all the cellular machinery they need to generate energy and reproduce. Some carbon is incorporated into long, stiff polymers of cellulose and lignin for the fortification plants need to stand tall and defend against intruders. And then there is that remarkable habit of plants—they carry out the very expensive process of photosynthesis and then dump up to one-third of the fixed carbon into the soil around their roots.[10]

The region around roots, known as the rhizosphere, is a smorgasbord for microorganisms. A cascade of diverse chemicals gushing from the root differentiate this land of plenty from the relative poverty of the rest of the soil. Bacteria swim around the root, gobbling up the goodies. Some bacteria attach to the root, making it look like a spiked medieval mace, getting in line to taste the delicacies secreted by their host. They can then convert this food into the glue that sticks together small particles to form the aggregates and clods that typify healthy soil structure.[11] The rhizosphere microbes create a phalanx around the root that protects it from unwelcome invaders and pathogens also attracted to the food oasis. Plants provide a steady stream of nutrients to their microbial denizens who return the favor by building soil structure and arming their hosts against marauders.

It's not just bacteria that exchange plants' carbon for other nutrients. Mycorrhizal fungi, for example, have been collaborating with plants for four hundred million years. Today, 92 percent of plant families each form intimate associations with one of the thousands of mycorrhizal species living in soil. The fungi first infect roots, then create a fuzzy matrix of hyphae around the root that captures nutrients. Fungal hyphae are long, skinny cellular tubes that extend into the surrounding soil, where they solubilize phosphorus and other nutrients that are otherwise inaccessible to the plant. This symbiotic relationship reduces the need for phosphorus fertilizer—an agricultural amendment that may be hard to find within the next few decades as supplies of phosphate rock dwindle across the globe.[12]

As plant, animal, and microbial members of the soil community die off as a result of disease, seasonal cycles, nutrient limitation, or old age, microorganisms decompose their remains, returning the complex molecules to simpler building blocks that can be reused for new life. Fungi in particular have tremendous degradative capacity, including the ability to pry apart cellulose and lignin—polymers so tough that few organisms can degrade them, which is why they are considered undigestible fiber in the human diet.

Over time, decomposing biologic material becomes organic matter, the organic component that differentiates fertile from infertile soils. The accumulation of organic matter is accompanied by improved nutrient and water retention and erosion and compaction resistance. Higher soil organic matter fosters higher biodiversity, which in turn reduces plant disease. Organic matter imparts a black or dark brown pigment to the upper soil layers as melanin—the molecule that gives human skin its hue—is released from decomposing soil microbes and insects. As organic matter accretes, so, too, does melanin, darkening the color of the soil, making black soil synonymous with fertile soil.

Soil organic matter benefits the entire planet by providing the largest terrestrial store of carbon. Soil around the world contains an estimated 2,500 gigatonnes of carbon, more than the total amount released by human activities since 1750 and the start of industrialization.[13] Soil management, therefore, has global implications for both food security and climate regulation (fig. 3). A solemn responsibility.

When raindrops land on soil, they disperse but don't stop. Many water molecules will bind to soil substances or get sucked up by plant roots and soil pores. Other molecules succumb to gravity and take the torturous path through layers of sand, silt, clay, organic matter, gravel, and bedrock into groundwater, or aquifers, flowing through porous rocks across continental crust.

Groundwater makes up 75 percent of the Earth's freshwater and accounts for approximately 40 percent of water used for irrigation and at least 50 percent of drinking water worldwide. It's extraordinary to think that, even though freshwater makes up less than 1 percent of all water on Earth (most is salty ocean), groundwater meets the daily needs of 2.5 billion people. About 700 million people depend on groundwater in India, making it the world's primary user of groundwater and a country where overpumping of wells has contributed to a water scarcity crisis. Collectively, India, China, Bangladesh, Nepal, and Pakistan use approximately half of the groundwater available worldwide. Their expanding demands have resulted in crises of depletion across the Asia-Pacific region. In Canada and the United States, about one-third of all citizens who are served with public water are supplied by well systems for which soil provides the main (and sometimes only) filtration step.[14]

Water filtration is perhaps soil's most overlooked service. People who would never sip from a muddy puddle happily drink groundwater that has been pumped from deep below that puddle. So, what happens between the puddle and the groundwater? Biological and chemical contaminants are removed by soil. Some chemicals adhere to soil particles, and others are degraded by the microbial workforce.

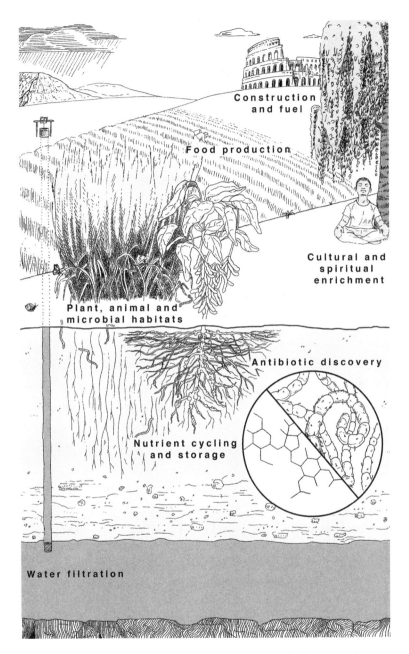

Figure 3. The diverse and vital services provided by soil. Illustration by Sophie Wolfson.

Collectively, soil microbial communities have an unparalleled metabolic virtuosity, as each microbe is endowed with distinct metabolic utilities. While some members are breaking apart herbicides, others are eating pharmaceutical compounds with an agility that chemists dream of.

Nowhere is this virtuosity more striking than in the degradation of gasoline. There are millions of underground gasoline storage tanks around the world, and most are leaking. For humans, this is a problem. For microbes, it's a picnic. Consortia of microbes tear gasoline apart and salvage the bits for their own growth and reproduction. If these microbial communities disappeared, many people would be drinking groundwater spiked with petrol.[15]

But soil does an imperfect job of purifying water. In general, the more chemicals are spread on the land, the greater the challenge for underlying microbes to keep pace, and the more likely that some toxic molecules leach into groundwater. Soil and rock can even contribute contaminants to water as it passes through, dissolving minerals and carrying them downward. When groundwater is pumped to the surface, contaminants can be detected by the smell of hydrogen sulfide, stains left by iron oxides, scale from calcium phosphate, and the taste of chloride salts.

On the flip side, soil is excellent at filtering out pathogens, which is why manure is usually a safe fertilizer. Microbes that are harmful to humans but harmlessly hosted by farm animals are found in abundance on land treated with manure, yet most do not reach the groundwater. Some pathogens die off in soil due to intolerable temperature, salinity, or acidity, whereas others are eaten or killed by competing soil microbes. Occasionally pathogens slip past the obstacles en route from surface to groundwater and cause outbreaks of illness. Though rare, these events are on the rise, in part due to feces-contaminated well water. *Legionella,* for example, the causal agent of Legionnaires' disease, appeared in groundwater in 2016 following extreme flooding events in Louisiana. As the global population continues to grow by approximately eighty million people each year, the value of freshwater cannot be overestimated, and soil's role as the world's largest water filter should not be overlooked.[16]

For millennia, humans have relied on soil microbes without understanding them. Ways to see these invisible creatures came in leaps. In the seventeenth century, Antonie van Leeuwenhoek created a lens that enabled him to see bacteria. In the nineteenth century, Robert Koch cultured bacteria on a solid medium, allowing growth of pure cultures of a single species. After a century of basing most microbiological discovery on microscopy and culturing, Norman Pace discovered a way to study the uncultured microbial world. To the surprise of many microbiologists, lots of bacteria do not grow on standard culture media. Environments thought to be sterile turned out to be teeming with microbes. Microbial communities thought to be simple turned out to harbor thousands of previously unknown species. Soil presented a cornucopia—fewer than 1 percent of its bacterial species are readily culturable, and the unculturable ones span a staggering diversity.[17]

Pace's method was based on the discovery of Carl Woese that molecular methods created a new window on evolutionary relationships, building a new Tree of Life that has guided biology ever since. The modern Tree of Life organizes life into three domains: Bacteria, Archaea, and Eukarya. Bacteria and Archaea descended from a common ancestor approximately 3.5 billion years ago, and Archaea and Eukarya split later (fig. 4). Although Bacteria and Archaea look similar—members of both domains are microscopic, unicellular, and contain

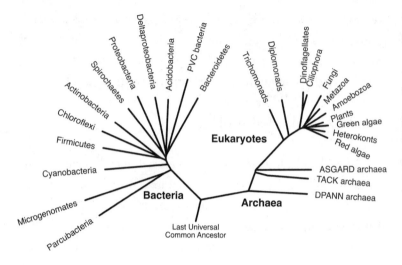

Figure 4. A phylogenetic tree of life constructed using the universal chronometer, 16S rRNA gene sequences. Illustration by Marc G. Chevrette.

no nuclei or other organelles—Archaea are, surprisingly, more closely related to the eukaryotes, the so-called higher organisms that contain nuclei.

It is immediately evident on gazing at the Tree of Life that the vast majority of life is microbial. All of the Archaea and Bacteria, and most of the Eukarya, are single-celled. And nowhere is the diversity of microbes greater than in the soil.

The diversity of microorganisms in soil has provided what might be the greatest boon to human health in our history—antibiotics. After the first antibiotic, penicillin, was discovered from a mold, it turned out that a real treasure trove lay among soil bacteria. From the 1940s to the 1980s, microbiologists and pharmaceutical researchers discovered a stupendous array of antibiotics produced by soil bacteria. These compounds became the backbone of the antibiotic industry, generating such drugs as tetracycline, vancomycin, and strep-tomycin. The gifts of the soil truly changed the course of human existence, removing such diseases as typhoid fever, tuberculosis, and typhus from the top ten causes of death in countries with access to antibiotics and making bacterial infectious disease treatable rather than deadly. Life expectancy rose from forty-seven years to seventy-nine years in the United States and other antibiotic-rich countries.

The prolific soil bacteria gave rise to a massive and profitable antibiotic industry whose products became commonplace in modern medical practice. But in addition to their use for bacterial diseases, antibiotics began to be prescribed for viral diseases, which they do not control, and became a widespread additive to animal feed because they have the surprising effect of accelerating weight gain in chickens, pigs, and cows. Over the past few decades, between 70 percent and 80 percent of the tonnage of antibiotics used worldwide has been fed to healthy livestock.[18] Many antibiotics do not degrade in the animals' bodies, so they wash into soil, groundwater, lakes, and rivers, affecting the communities of bacteria that live and work in those environments. The more we use antibiotics, the more resistant bacteria evolve and spread. Today that resistance threatens to return bacterial infectious diseases to the top-ten list of deadly diseases as routine infections are increasingly untreatable and fatal.

In 1985, my mother developed a bacterial lung infection. I had no doubt that she would take antibiotics and bounce back to her normal self. After all, she was a strong, brilliant woman who was as beautiful as her name, Blossom. She came from a family of tough Hungarian immigrants and a mother who was one of the original Rosie-the-Riveters who built fighter planes at Grumman Aircraft during World War II. As a bacteriologist, I had great faith in the power of antibiotics, so it was no surprise to me when a course of tetracycline restored my mother's health.

During the 1980s, the pharmaceutical industry began to abandon antibiotic discovery despite evidence that many pathogens had evolved resistance to existing antibiotics. The industry focused, instead, on developing drugs for such long-term conditions as high cholesterol and depression. From a financial standpoint, the choice made sense. Drugs administered over a lifetime are more lucrative than antibiotics typically taken for a few days. Arguing that the infectious disease problem was solved and that there were no antibiotics left to discover in soil, by the 1990s the industry had largely abandoned antibiotic discovery, and the pipeline of new antibiotics slowed to a trickle. In the 1980s, forty new antibiotics were registered for clinical use. In the 2010s, eight were registered. At the turn of the millennium, highly resistant strains of several major pathogens had spread across the globe. Many were resistant to multiple antibiotics, sometimes leaving no treatment options.

After my mother's first success with tetracycline, the infection recurred. To my family's relief, the prescribed antibiotics worked, and she returned to health. But then the infection returned. It returned again and again. She cycled through many antibiotics administered orally, intravenously, or by inhalation, but after several years her weakened lungs became infected with the virulent bacterium *Pseudomonas aeruginosa,* and it remained in permanent residence.

Meanwhile my own university research lab was discovering antibiotics, first by accident and then by design. It was becoming clear that the pharmaceutical companies were wrong—soil had *not* been tapped out. Rather, it had been abandoned. Driven by my personal sense of urgency, my team analyzed thousands of soil bacteria for antibiotic activity, hoping to find antibiotics that could save my mother. All the while, the *Pseudomonas aeruginosa* was developing resistance to drug after drug, impairing my mother's lungs and ability to breathe. I lost the race. Eventually, *Pseudomonas aeruginosa* developed resistance to all available drugs, and in 2001, my strong, beautiful mother lost the battle.

Today, there aren't many more antibiotics than there were in 2001. Interest in antibiotics has not resurged in industry, chemical and molecular methods have not provided a heap of new drugs, and cultured soil bacteria remain the best source of antibiotics. So to begin filling the gap left by departed pharmaceutical companies, I developed a worldwide network of college students working to discover new antibiotics in soil. My collaborators and I developed a curriculum called Tiny Earth that is now taught in twenty-seven countries to more than ten thousand students each year.[19] The students collect samples of soil and find antibiotic-producing bacteria living within them. It's my hope that this student workforce can replenish the pipeline of antibiotics and simultaneously develop an appreciation for the gifts of soil. On a personal note, Tiny Earth has turned my mother's defeat by pathogenic bacteria into a collaboration with the masterful soil bacteria.

Losing soil presents a problem for the future of antibiotic discovery. Some antibiotics have been found in only one soil sample, despite the many thousands explored.[20] So how many antibiotics will never be found in the 36 billion tonnes of soil lost each year?[21] As antibiotic-resistant bacteria threaten to return us to the preantibiotic era of medicine, when routine childhood disease, surgery, and an everyday cut finger could easily turn fatal, we need soil bacteria as partners.

Soil makes life on Earth possible for people. And yet we denigrate it by calling it dirt, bulldoze it without hesitation, and diminish its health with damaging agricultural practices. When we lose soil, we threaten our supply of food, clean drinking water, and new drugs. So why are we letting it erode at unsustainable rates?

4

Chaos to Orders: A Short Interlude

Georgianna Scott stood in a soil pit—a 2-meter-deep, square hole in the ground with tidy corners and a flat bottom. She faced the wall of soil, gazing up, then down, trying to make sense of what she was seeing. The clock was ticking. Within the hour, she would have to hand in her score cards to the judges who would select the individual champion of the 2018 International Soil Judging Contest held in Seropédica, Brazil. Few people have the chance even to watch the sport of soil judging, and here was a twenty-four-year-old from Clemson University in South Carolina who had come to Brazil to compete for the world title. Georgianna noted tangles of roots near the soil surface, a layer beneath that was pale and sandy, and an abrupt shift from sand to a clay so dense that her knife would not penetrate the pit wall. Below the clay she observed a red layer that became darker with depth except for white splotches here and there. This was unlike any soil Georgianna Scott had ever seen. What was the name of this soil?

We use one word, "soil," to describe myriad versions of the same basic mixture of rocks, water, life, time, and space.[1] Each dimension varies on many scales, creating the endlessly varied landscapes of the planet. Some features are shared among all soils, and others differ between continents or even between soil particles.

The human mind seeks patterns, imposing order on chaotic systems. Human and animal survival is predicated on pattern recognition to accelerate interpretation of the information around us.[2] We categorize things based on their similarities and differences, constructing taxonomies of living things, sounds, ideas, and objects. But we're most familiar with categories of things above ground. Although it is largely out of sight, the chaotic world beneath the Earth's surface can also be organized based on characteristics that can be observed today and serve as clues about soil's history and genesis.

Soil diversity starts with parent material. First, there's the underlying bedrock, such as granite, limestone, sandstone, schist, basalt, or shale. Minerals can also be added by wind or water that deliver materials from elsewhere, creating further complexity. Weather acts on the mineral mixture, carving out the physical and chemical environment that organisms will colonize and change further. As thermal changes fracture rock, the topology of the land influences the movement of water, which carries soil particles to new places, thereby altering the landscape's profile and the weathering process. Water and oxygen react with mineral surfaces, changing the chemistry of the parent material.

Millions of species of microbes, plants, and animals contribute to the chemical and physical processes that convert rocks into vast varieties of soil habitats across the globe. Roots chisel through the maze of particles from surface to substrate, each species and root type forging a different path. Large roots penetrate aggregates and create pores, whereas fine roots bend to the soil's forces, growing around aggregates and through existing pores. All roots leave their signature by contributing an immense amount of biomass to soil—the total fine-root structure of a single rye plant measured 620 kilometers, and the deepest root found to date, belonging to a fig tree in South Africa, extended 160 meters below ground.[3] Both the type of vegetation and uses of soil affect its development—a forest soil is influenced by the trees and understory and differs from a soybean field or pasture, even if they have a common geologic base.

The interplay of geology, vegetation, and weathering influences the tens of thousands of microbial species found in each gram of soil, further expanding the Earth's subterranean repertoire. Microbial secretions generate clumps, defining the characteristic channels and pockets that form patterns unique to each soil. The microbes also alter the chemical states of minerals, speeding up chemical reactions by as much as one thousandfold. The chemical changes contribute to the colorful underground display of soil layers, known as horizons.

Time influences soil, varying across an enormous scale, from seconds to billions of years. Rocks change on a millennial schedule, plants and animals typically live and die in month- to year-long increments, and microbes can multiply in minutes or lie dormant for millennia. Superimposed on these changes are rapid shifts in soil temperature, water, and air driven by daily cycles and weather events. Trees take up water during the day, and some unload great volumes of it into the soil at night. A rainstorm can transform parched land into a flood zone within minutes. All of these processes are in flux and at different stages, even in a single plot of land, so classification of soil is based on a snapshot in time.

Soil varies spatially on tiny and vast scales. Heterogeneity is its hallmark—soil is by definition a mixture of many types of particles, organisms, and chemicals, each in many phases of their life cycles. Microbial communities vary from clod to clod, particle to particle, and even across different locations on the same sand grain. In a clump of soil, the inner crevices may be free of oxygen, whereas the surface may have easy access to air. At the field scale, biological features vary most, with only small differences in the physical features unless the field straddles a geologic transition or contains a depression that accumulates water in one section. Spanning larger land areas, soil varies in all features as the underlying geology, weather, flora, fauna, and microbes change across the landscape.

So how do soil scientists make sense of this tumult of rocks, water, and life to describe a particular soil? What kind of taxonomy could bring order to the chaos that is Earth's dark matter? We need a language, a shorthand, to quickly communicate the features of a soil—to conjure the image of a black soil in Ukraine or a red soil in Alabama with a single name. That is the science of soil classification and the art of soil judging.

If we joined Georgianna Scott in a soil pit, we might first be struck by the bouquet of geosmin. We could sample a bit of soil and taste it for texture, measuring the size of the particles—gravelly, sandy, or as smooth as talcum powder. It might taste sharp and acidic or sweet and chalky. We might see dramatic strata or subtle layers. Wavy horizontal stripes of black, brown, gold, red, or gray narrate the events, ancient and recent, that shaped this particular piece of earth. The roots near the surface would be embedded in a brown matrix, known as the O, or organic, horizon, that contains the lushest mixture of living and decaying animals, plants, and microorganisms of any soil layer. Working down the soil profile, next is the A horizon, influenced by the life-forms at the surface but distinguished by more complete decomposition. A typical soil might contain below the A, in order, the B horizon or subsoil (present only in some soils), the C horizon containing parent material, and the R horizon of bedrock.[4] Some soils are marked by an E horizon, lying below A, which has been depleted of clay, many minerals, and organic matter, leaving behind only the most resistant minerals. The presence, thickness, and color of the horizons bear witness to how the living things and rocky material have been molded by exposure to water, weather, time, and one another. The features we see before us are clues to the classification used by many soil judges like Georgianna as they fill in their score cards (fig. 5).

The first people to classify soil were farmers. Ever since farming began, roughly twelve thousand years ago, people have differentiated soils based on relative productivity. They passed down wisdom from abundant and failed harvests to descendants who enriched soil folk knowledge with results from experiments of their own. After farmers, it was government officials of ancient civilizations who advanced soil classification for economic gain. In China around four thousand years ago rulers of the Yao dynasty (2357–2261 BCE) divided arable land into nine classes based on crop yield and taxed landowners accordingly.[5]

A call for soil classification in 1832 was answered by two geologists in the late nineteenth century—E. W. Hilgard in the United States and Vasily Dokuchaev in Russia—who developed ways to group soils based on their properties. Both were way ahead of their time.

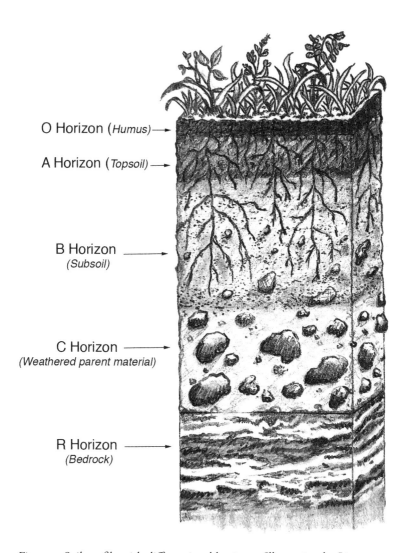

O Horizon (*Humus*) —→

A Horizon (*Topsoil*) —→

B Horizon ———→
(*Subsoil*)

C Horizon ———→
(*Weathered parent material*)

R Horizon ———→
(*Bedrock*)

Figure 5. Soil profile with differentiated horizons. Illustration by Liz Edwards.

Hilgard's classification ideas were not adopted for another fifty years, and his writings about the relationship of soil and climate are reflected in twenty-first-century thinking. Dokuchaev was the architect of the nineteenth-century Russian school of soil science, whose tenets scaffold today's thinking. He recognized soil as a unique part of Earth and integrated it into a unified vision of the natural world. Dokuchaev observed soil horizonation, the basis for his "genetic soil classification," which is based on the factors that led to formation of the soil, or the process of soil genesis. His description of soil as the "result of the totality of activity of the bedrock, the living and inanimate organisms (plant as well as animal), climate, the age of the country, and the topography of the surroundings," closely matches our modern understanding of soil.[6] Dokuchaev had a lasting effect on the field that he created, coining Russian terms still used throughout soil science. Fittingly, a crater on Mars is named for him.

In the early twentieth century, U.S. scientists surveyed the country's expansive land in the hope of splitting its soils into groups. Soil scientists working in the 1920s began seeing how distinct processes formed different soils, as Dokuchaev had half a century earlier. Leaching (water-instigated removal of soluble nutrients), melanization (incorporation of organic matter), and oxidation (reaction of minerals with oxygen) created horizons that were easily differentiated by color. The emphasis on productivity and soil genesis diminished, replaced by the goal of capturing a wide range of characteristics—texture, chemistry, and depth—in a classification system.

Soil classification took off after World War II when the Soviet Union, France, the United States, and several other countries commissioned scientists to develop national classification systems. Soviet scientists built on the nineteenth-century work of Dokuchaev, sorting soils based on the processes that formed them. In France, a 1967 soil commission grouped soils by shared characteristics (degree of wetness, for example) into classes, subclasses, groups, and subgroups. These two soil languages flowed into other territories along political fault lines. The French method was adopted by former French colonies throughout Africa, and the disparate systems used in East and West Germany were reconciled only after the Berlin Wall fell. China originally classified its soil using the U.S. paradigm, transitioned to the Soviet system in 1949, and now uses its own lexicon. Australia, Brazil, Canada, England, Wales, New Zealand, South Africa, and several other countries use their own systems. The United Nations Food and Agriculture Organization attempted to broker a universal language, the World Reference Base for Soil Resources, now used by half the world's countries, but it never became truly universal.[7]

The Soil Taxonomy system is used by the other half of the world, including much of Africa and the Middle East, South America, India, South Asia, and the United States. Soil Taxonomy is based on soil horizons' properties, including texture, weathering, and thickness. The system has twelve orders, the broadest divisions, which convey the salient features of each group's morphology. The most fine-grained groupings are known as soil series, and there are more than twenty-one thousand soil series—or species of soil—in the United States.[8] In between orders and series, there are categories based on the weather the soils have experienced and the kind of work they do.

The twelve soil orders of the U.S. Soil Taxonomy system capture information such as soil texture, pH, and parent materials. The Gelisols and Aridisols, for example, are named for such characteristics as their ice-veined horizons and low organic content, respectively, that reflect the climate in which they are found (figs. 6, 7). Names of other orders are based on visible features of soils' horizons, which also reveal elements of their genesis. The Entisols and Inceptisols are the youngest soils, in which distinct horizons have not yet formed. Oxisols and Ultisols are older, highly weathered soils that are usually

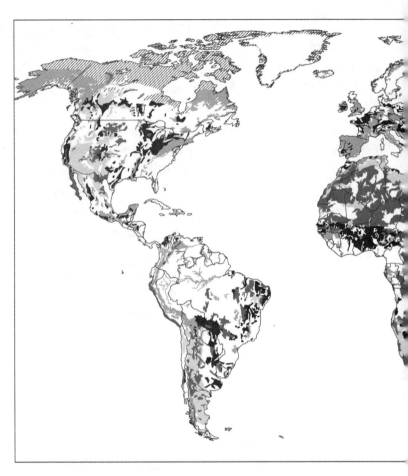

Figure 6. World map of the most abundant soil orders, featuring Entisols, Inceptisols, Gelisols, Aridisols, and Alfisols. Illustration by Bill Nelson and based on FAO-UNESCO, Soil Map of the World, digitized by ESRI; and Soil Climate Map, USDA-NRCS, Soil Science Division, World Soil Resources, Washington, D.C., April 1997, rev. September 2005.

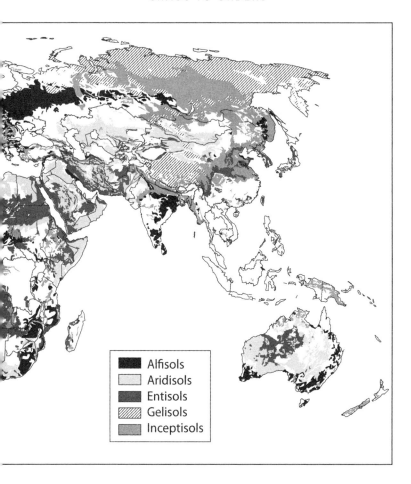

found in the tropics as well as Spodosols, found in temperate zones, often sport brilliant colors (see pl. 1). The twelve orders are as follows:

Alfisols (10 percent global ice-free land, or GIL) are fertile soils used for agriculture and forestry. They form in humid climates beneath deciduous forests or in semiarid regions beneath savanna vegetation and take thousands of years to develop, with many over four

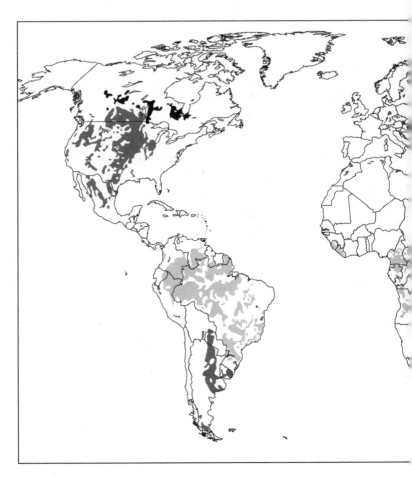

Figure 7. World map of Mollisols, Histosols, and Oxisols. Illustration by Bill Nelson and based on FAO-UNESCO, Soil Map of the World, digitized by ESRI; and Soil Climate Map, USDA-NRCS, Soil Science Division, World Soil Resources, Washington, D.C., April 1997, rev. September 2005.

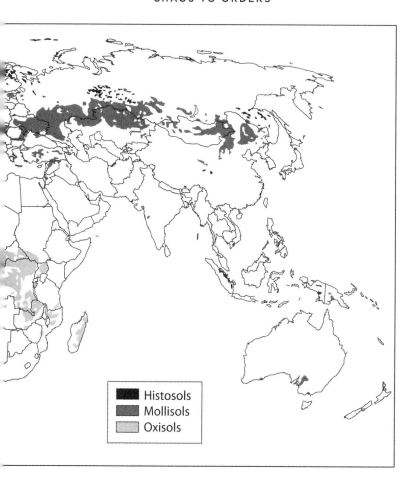

Histosols
Mollisols
Oxisols

hundred thousand years old. Rainfall transports clay through the Alfisol profile, bleaching the upper horizons.

Andisols (1 percent GIL) are mostly acidic yet fertile. They emerge from the ash of erupted volcanos and store high amounts of carbon. Andisols support as much as 10 percent of the human population worldwide as well as diverse plant species and are typically found in humid tropical regions with high rainfall.

Aridisols (12 percent GIL) represent one-third of the Earth's deserts and are mostly found in Asia, Africa, and Australia. They are among the oldest but also shallowest soils, storing little organic matter and the least carbon and nitrogen among all soil orders. Aridisols support agriculture only when supplemented with nutrients and water.

Entisols (16 percent GIL) are the most abundant soils on Earth and are typically young or derived from weathering-resistant parent material, resulting in minimal horizon differentiation. Entisols underlaid ancient agricultural hubs along the Nile and Tigris Rivers.

Gelisols (13 percent GIL) form in cold climates and contain a layer of permafrost within 1 meter of the surface. Frequent freezing and thawing events create Gelisols' irregularly shaped horizons veined with ice. These soils provide food, shelter, and fuel to many indigenous communities and sequester roughly one quarter of all organic carbon in soil.

Histosols (1 percent GIL) are wet and rich in organic matter. They are found mainly in boreal forests in peatlands, mires, bogs, fenlands, and swamps. Burned for fuel and added to other soils for fertility, Histosols also provide such ecological services as flood control, wildlife habitat, groundwater recharge, carbon storage, and nutrient cycling.

Inceptisols (15 percent GIL) are typically young and barely differentiated from the underlying parent rock. They occur along sloped landscapes and fluvial plains, containing deposits from eroded mountaintops and water features. Used for forestry, pastureland, and agriculture, Inceptisols support 20 percent of the world's population—more than any other soil order.

Mollisols (7 percent GIL) or *black earths* are deep and fertile, carrying a thick top layer of organic matter. Formed beneath grassland and enriched by wind-blown dust as well as sediments from glacial drift, this soil order is prevalent in Ukraine, Russia, northeastern China, Argentina, and the United States. Mollisols are among the most agriculturally productive soils and are used for growing wheat, soybean, corn, and millet. Mollisols are good at sequestering carbon. Prairie plants that gave rise to Mollisols transport copious organic carbon into their roots, producing more biomass below ground than above.

Oxisols (7.5 percent GIL) are red or yellow and iron-rich. They form in the tropics, mostly in South America and Africa. They can develop in situ, slowly transforming weather-resistant parent material, or from soil deposited from elsewhere. Effectively infertile, these soils support plant growth only when an overlying forest is burned to release essential nutrients or through the application of fertilizers.

Spodosols (2.5 percent GIL) or *white earths* are typical of cold regions with heavy rainfall such as the coniferous forests of Canada, Russia, Scandinavia, and many mountain ranges. There are also large expanses of Spodosols in the tropics. Formed by leaching organic matter, aluminum, iron, and silicon through the profile, Spodosols have a starkly white E horizon and can be rich in carbon. Spodosols take at least three thousand to eight thousand years to develop and are often too acidic and sandy to support crops other than potatoes, apples, barley, and berries.

Ultisols (8 percent GIL) mostly form beneath tropical forests. Their distinct horizons can take millions of years to develop. They are relatively low in available nutrients, which are largely stored in the lush plant life of the forests.

Vertisols (2.5 percent GIL) are dark, clay-rich soils that form atop all kinds of parent rocks and under a range of climatic conditions, especially monsoons. If water is managed well, Vertisols can be highly productive for crop and pasture agriculture, though dry-season cracks pose a danger to animals.[9]

Soil classification acquired a new cachet when the International Soil Judging Contest started. For Georgianna Scott, it was a road to rapid fame in the soil science community. Within a year of starting to judge, she had successes at regional, national, and international competitions that came as successive shock waves to her and the community. But in retrospect, signs that she would excel were evident from childhood. For her fourth-grade science fair project, Georgianna dug a hole in the woods just to see what was there. She found earthworms, leaf litter, and the jackpot: an arrowhead, sleek and white, almost 8 centimeters long, an inheritance from the Catawba Indians who had once lived on the land.

More than a decade later, standing in the soil pit in Brazil, Georgianna struggled to identify the soil's order and suborder. It didn't seem to fit any of the twelve orders in the U.S. Soil Taxonomy. Her mind returned to her coach's advice not to jump to conclusions but to simply observe and record. She made meticulous observations: the soil had a 20-centimeter A horizon, where root systems were evident in various stages of decomposition; its pale, sandy layer had probably lost its minerals through leaching. The white patches in the red layer were likely caused by water saturation. It was the sharp demarcation and sudden change in texture between the horizons that tipped her off. Fortunately, Georgianna had learned both the U.S. Soil Taxonomy and the World Reference Base classification and was reminded of the major WRB soil known as a Planosol, typified by abrupt textural changes between horizons. Using her astute observations and knowledge of both soil classification systems, Georgianna won the competition and was crowned the individual champion of the 2018 International Soil Judging Contest.

Soil names aren't just labels; they tell stories. Take, for example, the great soil order Mollisol, which is known for its beauty and utility. Mollisols are deep, black, and fertile and among the most productive soils in the world. Mollisols were first named by Dokuchaev as *Chernozem,* meaning black earth in Russian, another name still used today. These soils are typically found in prairie and steppe regions of temperate environments such as Ukraine and the American Midwest. In these plains, organic matter is deposited by deep-rooted perennial plants and herds of large animals roaming the land. Over centuries, the plants and animals—with lots of help from the soil microbes—create fertile, dark topsoil. That's a lot of information packed into a name.

Entisols are the youngsters. Many haven't had time to differentiate horizons from parent material either because the substrate just formed or because a necessary ingredient for weathering the soil, such as water, is lacking. Soils can also be kept perpetually young by continual water saturation, which slows the differentiation process by preventing oxygen from penetrating the soil.

In the great deserts of the world such as the Sahara, Gobi, and Mojave, we find the Aridisols, which are dry, arid, and fragile. They support a surprising diversity of plants, animals, and microbes but are vulnerable to mistreatment by overgrazing livestock, for example, and can rapidly lose their ability to support life. The Aridisols are typically rich in minerals that have not been removed by water filtration, so when sufficient water is present, these arid soils can be quite productive. They also contribute to the productivity of soils far away as wind transports billions of tonnes of Aridisols from one continent to another each year.

It is captivating to look at an exposed soil profile and imagine its origins and pathway to the present. For example, I live on a line delineated by the last glaciation event in Wisconsin. On the east side of the line is a terminal moraine where the colossal slab of ice stopped, leaving behind deep deposits of rocks, gravel, and debris. On the west and unglaciated side of the dividing line, the soil is silty and soft. The border of the two affords a stark glimpse of events that occurred ten thousand years ago.

South of the glaciated regions, we find very different soils, including the Ultisols, which occur in warm, humid climates around the world. Silica, iron, and clay typically leach from the upper layers of these soils, bleaching them to gray or white. The particles are deposited in the lower soil profile, where they react with oxygen, forming the soft reddish or yellow color typical of metal oxides.

All of the Earth's soils tell a story of the past. It's up to us to write their future.

5

Wind, Water, and Plows

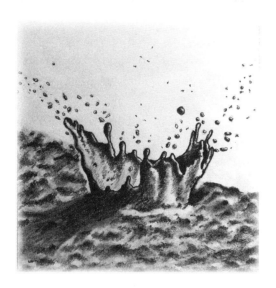

Today farmers around the world are grappling with erosion (see pl. 2). Instead of fertile O and A horizons, where the fertility and life largely reside, some farmers are left with only the subsoil or, worse, the rubble of rocky parent material. Others are just starting to see erosion nibble away at the topsoil and with it, productivity and profit.[1] For these farmers erosion is no abstract threat; it is a tangible loss of the primary resource upon which their livelihoods depend. Some are caught in a destructive cycle of land degradation, propping up yields with excessive fertilizer until erosion renders the land unproductive. Families are forced to abandon exhausted land and find new ground to grow their crops and feed their livestock. But erosion is not an inevitable outcome of farming. Many farmers maintain healthy soil, restoring the nutrients their crops and animals remove and preventing soil displacement. Some indigenous people have been doing this for centuries, and other communities are just discovering soil conservation practices. To guard soil for future food security and environmental stability, we need to understand good soil stewardship and adopt its practices more broadly. But first, it is worth delving into the nature of erosion. Let's start with its causes.

Erosion is a natural process caused by wind and water. It is as old as soil itself, chiseling the Earth's surface to its current splendor, creating the sinuous paths taken by rivers, revealing craggy outcroppings of some geologic relics and molding the graceful contours of others. Although its impact can be good or bad, erosion itself is neutral. When soil particles are detached and moved to a new location, we call it erosion. We often refer to soil "loss," but in reality, erosion means that soil is lost only from its original location. This usually means removal from agricultural settings where it is needed.

The impact on the places where soil lands can be either ruinous or enriching. Fertile soil may end up where it is useless to agriculture—in ditches, on roads, or buried under less fertile layers where it is inaccessible. Some settles in reservoirs, reducing their water-holding capacity by as much as 50 percent, and some clogs waterways, where its nutrients can be liberated to support overgrowth of hungry microbes that destabilize local ecosystems and suffocate aquatic life.[2]

In contrast, some deposition events are a boon to the local environment because they replenish and enrich local soil, delivering much-needed minerals to nearby fields or distant continents. This is an annual occurrence around rivers that flood their banks, distributing upstream soil onto adjoining land. The great deltas around rivers such as the Nile and the Mississippi form when rivers recede, producing some of the most productive agricultural land in the world.

If it has always occurred, what's the problem? Why are current rates of erosion troubling? At any location, soil depth is determined by the balance of soil genesis and deposition with the opposing force of erosion. The natural rate of topsoil production is at most 0.5–1.0 tonnes of topsoil per hectare per year. With a global annual erosion rate averaging 13.5 tonnes per hectare, there's a problem—*soil is vanishing from the locations where it was produced on average ten to thirty times faster than it is generated.* That's not sustainable! In many regions of the world, farming practices have accelerated erosion to a pace that endangers food security. Erosion is hastened further by climate changes that bring aggressive rainstorms and climbing temperatures that compound the effects of farming on soil. As the human population grows, farming intensifies, climate change worsens, and soil is the victim.[3]

The effects of unsustainable erosion are felt at every level, from personal to global. On farmland, erosion reduces crop yields and leads to more erosion, exacerbating farmers' financial hardship that follows. Erosion of some soils, especially peaty Histosols, often leads to transformation of soil carbon to greenhouse gases that warm the planet. Erosion's role in the global carbon budget is understudied and remains controversial, but as a consequence of the production of greenhouse gases that sometimes follows erosion, it may contribute as much as 2 gigatonnes of carbon to the atmosphere annually.[4] That represents around 20 percent of annual global emissions from burning fossil fuels. Some compensation for these gases is probably gained from eroded soil that is buried where it is deposited, reducing its potential for release of greenhouse gases.

What are the natural drivers of soil erosion? Erosion by wind is perhaps more famous than by water because windstorms dramatically obscure visibility and create patterns that can be observed from Earth or in images procured by satellites. Dry soil is especially susceptible to wind because dry particles can be dislodged more easily than those anchored by a film of water. Wind-generated forces, known as aeolian processes, deplete the contributing land and drive formation and enrichment of recipient soils around the world. Among the most vulnerable to erosion are the arid and semiarid soils, such as Aridisols and some Entisols, that cover 40 percent, or 430 million hectares, of the Earth's surface. The annual global dust emissions from drylands and deserts are somewhere between 1 and 4 gigatonnes, with more than half from North Africa. Winds whisk soil particles off the surface of the Sahara Desert and carry them far above the Earth to distant parts of the world. Incoming desert soil enriches phosphorus stores in South America's Amazon rain forest. Similarly, dust from Asia sustains Hawaiian land and can even reach the west coast of North America. The deposited minerals nourish American soil, but at the cost of depleting desert soils in Africa and China. Some of the same lands that receive soil gifts from other continents also suffer their own erosion. In the United States alone, wind erosion is estimated to remove 0.63 billion tonnes of soil annually, predominantly from dry croplands.[5]

Wind erosion events punctuate history. One of the worst dust storms in United States history struck on April 14, 1935. The date is remembered as Black Sunday because winds measuring 100 kilometers per hour tore through the Great Plains, sweeping almost 1 million tonnes of dry topsoil into the air, blotting out the sun in Oklahoma. Black Sunday marked the midpoint of the Dust Bowl decade, a drought spanning the 1930s, which was studded with windstorms that removed massive amounts of the mighty Mollisol, the foundation for agricultural productivity in the United States. Years of poor land management, such as heavy tilling and planting soil-depleting crops year after year, had left the Plains soil vulnerable to wind erosion.[6]

Wind erosion is the leading cause of desertification in northern China, turning fertile soil into arid land. Although wind erosion has sculpted much of China for centuries, and dust storms have been documented since 205 BCE, during the past seventy years soil loss has heightened, making the Gobi the fastest-growing desert in the world. More than 70 million hectares have been degraded, most of them now infertile. Close to 30 percent of China's desertified land is used for livestock and crop production, which amplify its vulnerability to erosion.[7] Affected farmers sometimes buttress crop production with high fertilizer application, but this is a temporary fix for yield, and it spawns environmental problems such as pollution of waterways and production of nitrous oxide—a potent greenhouse gas. Moreover, when farmers boost yields with excessive fertilizer that masks the impact of erosion, they may be less likely to address soil loss, which may be catastrophic for their soil in the longer term.

India is known for its regular dust storms, but in 2018 a particularly powerful storm ravaged the northern states of Uttar Pradesh and Rajasthan. High-velocity winds arrived before the rainy monsoon season began, so the soil was dry and erodible. Trees and utility poles were ripped from the ground and buildings fell, taking more than one hundred human lives. India's repeated dust storms, which cause erosion, air pollution, lung disease, eye damage, and loss of human life are thought to result from poor soil management practices in agriculture coupled with drought and severe winds (fig. 8). But as much as it erodes soil, wind is responsible for only 18 percent of the total erosion on India's impressive 180 million hectares of agricultural land; the rest is caused by water.[8]

Figure 8. Dust storm. Illustration by Helen Jones.

Compared with the dramatic effects of wind, erosion by water is often less visible but far more pervasive. In fact, water is the most prevalent earth mover worldwide, detaching particles from the bulk soil and propelling them into rivulets, gullies, and streams (see pls. 3 and 4). Water meets soil through flooding, irrigation, and rain events, which are particularly important because of the force with which raindrops hit the ground. Individual raindrops may seem gentle, but their collective power is literally earth-shattering: 100 centimeters of rain pummeling the surface of 10 hectares deliver the kinetic energy equivalent of 1 tonne of TNT explosives. To a clod of soil, rain can be cataclysmic. Worldwide, water is thought to remove 20–50 billion tonnes of soil annually from its original location.[9] This is expected to increase as climate change intensifies, bringing more severe rainstorms to locations around the world.[10]

Water erosion is most destructive on sloped land where gravity pulls water to lower elevations, sweeping the soil along with it. The steeper or longer the slope, the greater the potential for water erosion. The resultant changes to the landscape range from imperceptible to total remodeling, where small rills or great gullies serve as highways for soil moving across the landscape slowly or rapidly depending on slope, obstacles, and water depth (fig. 9). Soil eventually settles in lower positions on the landscape or continues on its journey into culverts, reservoirs, streams, rivers, and the sea, where it may help or hurt the local ecosystem.[11]

Water falls unevenly across the globe, but its erosive power does not discriminate. In Africa south of the Sahara, water erosion is estimated to have degraded 46 percent of the land, including 80 percent of Nigeria. On the volcanic islands of the South Pacific, the steep topography and intense storms conspire to cause average erosion of 50 tonnes of soil per hectare per year, which is exacerbated in Papua New Guinea and the Solomon Islands by deforestation that has left the soil vulnerable. In India, water plays many roles, good and bad. Although the country suffers from chronic water shortages, water has eroded soil from more than ninety million hectares, or one-third of India's total land. Between 1950 and 2008, India more than tripled its irrigated land area, which has boosted food production to unprecedented levels but has also damaged soil by salination—a process that brings salt dissolved in groundwater to the surface, where it accumulates. At high levels, these salts stunt plant growth, making the plants less supportive to soil production and ultimately heightening erodibility.[12] Water issues challenge India's ability to remain food self-sufficient, which will demand almost doubling agricultural production from 2006 levels to feed the population of 1.62 billion people projected to inhabit India by 2050. A growing population cannot be sustained easily on a shrinking fertile landmass.

Figure 9. Eroding rill on an intensively farmed field. Illustration by Liz Edwards and based on a photograph by Katharina Helming.

Critical to conclusions about erosion is the ability to estimate it accurately. Tracking the journey of soil from where it is dislodged to its resting place has never been an easy task. Farmers have undoubtedly observed soil move from their fields or accumulate elsewhere since the beginning of agriculture. Casual observations document erosion before it was studied formally. For instance, an investigation of Russia's Svir River in 1897 estimated that a meter of soil had been deposited in the river throughout the previous century based on the depth of hundred-year-old coins found buried in the river's sediment. But until the twentieth century erosion had scarcely been measured systematically. In 1915, Ray McClure, an undergraduate student at the University of Missouri, set out to study nutrient loss in runoff water from agricultural fields. While conducting his research, McClure asked his adviser how his measurements should handle the sediment that was carried in the runoff water from high to lower ground. The professor advised him to measure the sediment and nutrients in it, and McClure found that the nutrient levels in the runoff water and displaced soil were greater than the amount of fertilizer that had been applied to the fields, indicating a net loss of nutrients from the fields. His research also quantified the amount of soil lost from the fields, and thus began the study of soil erosion in the United States.[13] His investigation also illustrates the important property of erosion: displacement from a point of origin. In McClure's case, he could measure the accumulation of the soil down the hill from where it was dislodged. In many cases, the moved soil is as good as lost to its original landlord because it is buried in another field, spread across roads, blown to another continent, or washed into waterways.

Since McClure's experiments, the study of erosion has become more sophisticated. For decades, soil scientists have been estimating the rate of erosion from agricultural land using five methods. None is perfect, and each needs to be used with consideration of sufficient sampling and appropriate comparisons. They are:

Soil depth. We can measure the thickness of the surface layers of soil. Some researchers measure the depth from the surface to the soil's parent material. Others use the depth from the surface to the base of the A horizon, focusing on the portion richest in organic matter and excluding subsoil. The location of the subsoil and parent material doesn't change over time, so the distance from the surface to these layers indicates the amount of soil overlaying the mineral substrate. To estimate how much has been lost since the beginning of cultivation, we can compare the soil depth on an agricultural field against a nearby uncultivated area where the soil depth is likely similar to what it was when the land was converted to agriculture. Even better, repeated measurements of the same location can estimate soil loss over time. In a powerful application of soil depth comparisons, Jessica Veenstra and Lee Burras at Iowa State University assessed the impact of continuous row-crop agriculture over fifty years. They used data from a 1959 soil survey that described soil profiles at eighty-two sites representing twenty-one Iowa counties and compared them with profiles taken at the same locations in 2007. The thickness of the top horizon diminished on average from 15 to 1 centimeters across the eighty-two sites, and the hill bottoms accumulated soil. The eroded soil lost its aggregate structure during transport, rendering it far less healthy than it was in its original location. The study indicates that the top horizon shrank by 90 percent in forty-eight years. The plant roots were experiencing a very different, and less fertile, soil environment in 2007 than when the first samples were collected in 1959.[14]

Runoff or sediment. Soil erodes from one place to another, so we can estimate erosion based on the amount removed from one location or accumulated in the other. Researchers can estimate soil leaving a site by placing soil of a known weight in a mesh bag and measuring loss over time. By placing bags at enough locations across a field and validating the results with another method (such as runoff measurements), scientists can obtain a reasonable estimate of soil loss. To measure runoff at the point of deposition, they can install a vessel downslope from the field of interest to catch the soil that runs off the field.

Gutters or other containers can be used to collect the soil at intervals across the lower area. The weight of soil that accumulates in the vessel is measured over time, and averaging across collection units provides an estimate of soil erosion per hectare per time period. One study found that estimates of soil loss and downslope accumulation on the same fields were almost identical, indicating that these methods were likely measuring the same process and providing some confidence in the estimates. Long-term trends are measured through sediment accumulation in reservoirs and other bodies of water, but these measurements grossly underestimate erosion because less than half of eroded soil typically ends up in the water. Sediment collection in waterways is particularly powerful when used to compare the impact of management practices on the land that generates the sediment. In a study conducted by Wayne Erskine's group in New South Wales, Australia, researchers used a series of small dams to collect soil eroded from upslope land. The dams below cropland collected triple the amount of soil as those below forestland, indicating the erosive impact of cropland management. Even if the absolute amount of sediment collected underestimates erosion, comparing the relative quantities among different land management regimes is meaningful.[15]

Radioactive isotopes. Before the 1996 ban on nuclear testing, the United States, former Soviet Union, and several other countries conducted more than two thousand tests of nuclear weapons. Five hundred were nuclear bombs detonated aboveground that spewed radioactive byproducts into the atmosphere, some of which were eventually deposited on soil across the world. Similarly, the explosion in 1986 of the nuclear facility at Chernobyl in Ukraine and other nuclear accidents released radioactive fallout that settled on soil across the world. Soil particles rapidly bind radioactive elements, anchoring them in place and making the soil surface more radioactive than deeper layers. As the soil erodes from the surface, the radioactivity diminishes in its original location, providing an estimate of soil loss. Researchers have validated the radionuclide method by calibrating it with direct measurements of sediment accumulation and soil redistribution. The measurements are most useful when comparisons are made to baseline values at initial time points or compared with patterns of radioactivity on the surface of uncultivated land. The accumulation of radioactivity can also be used to determine the path of erosion and points of soil deposition. A second method, using radioactive beryllium, is used primarily for estimating erosion in geologic time or before human intervention. Radioactive forms of beryllium are relatively rare in the Earth's crust. They are generated by cosmic rays impinging on the Earth's surface and thus can be used to trace what was once surface soil in sedimentary rock layers.[16]

Remote sensing. The launching of the first satellite into space in 1957 gave us an entirely new view of our planet. Satellite images, or remote sensing, have proven useful for assessing soil moisture, roughness, vegetation, and topography. The images created from visible or infrared light reflected from the Earth's surface are used to assess features of soil. The characteristics that affect water infiltration into the land, such as presence or absence of a soil crust, porosity, moisture, and plant residue or canopy, can be detected on satellite images. The process of desertification has been monitored, and soil management practices, such as tillage (which affects the roughness) and cover cropping (based on exposure of bare soil) can also be observed. Large gullies can be directly detected, as can soil in waterways, because water's reflectance of visible and infrared light is increased by suspended sediment. The power of satellite imaging lies in the large land area that can be surveyed and how frequently it can be measured. The United States launched its first Landsat satellite in 1972 and its eighth in 2013. Landsat 8 orbits Earth every ninety-nine minutes at an altitude of 705 kilometers, and in sixteen days it captures the entire Earth's surface. Satellites also provide a historical record of soil erosion. In 2020 researchers assessed erosion in Slovakia over time using remote sensing images from 1949 to 2011.[17] Satellite data also democratize science—information gathered by Landsat is made available on three U.S. Geological Survey websites, so that eventually anyone will be able to track changes in soil.

Modeling. Soil scientists started correlating direct measurements of erosion with environmental factors in 1940 when the first quantitative relation between the slope of the land and erosion was established. The next association was between rainfall and erosion, established with measurements on approximately eight thousand plots. In 1965, several parameters were linked to erosion in a single mathematical model—the universal soil loss equation, or USLE, which computes the effects of rainfall, soil erodibility, slope length and gradient, and crop and soil management practices to estimate sheet and rill erosion (uniform removal of the surface layer of soil or transport of soil through channels, respectively). USLE was initially developed based on data from ten thousand measurements and has been validated with many more thousands of data points over the last six decades. The USDA's triennial Natural Resource Inventory reports soil erosion for every state using USLE. A significant limitation of USLE is its inability to account for erosion in gullies—the ephemeral deep cuts in land where masses of soil can wash away during storms; thus, in areas plagued by gully washes, USLE substantially underestimates erosion.[18] It is also limited by insufficient data in many locations, forcing the equation to use averages across large scales of time and space.

A second modeling approach, that of process-based estimates, is exemplified by the water erosion prediction project, or WEPP, which uses the processes of hydrology, plant growth, hydraulics, and erosion mechanics to integrate measurements of four factors—precipitation, topography, soil characteristics, and land use—into a sophisticated computer simulation of erosion. Whereas USLE correlations are based on static measurements that are correlated with past outcomes to estimate erosion in new settings, WEPP uses physical processes to integrate information into a prediction. Moreover, WEPP has the capacity to integrate satellite and other remotely sensed data collected frequently and at a granular spatial scale, providing a detailed portrait of continually changing weather and landscape. Integrating physical processes with rich data sets that capture the variation of terrain, weather, and farming practices across space and time makes WEPP erosion estimates applicable at many scales, including small fields to large watersheds. In an erosion project of unprecedented scale and accuracy, Professor Rick Cruse, with Research Manager Brian Gelder and their team at Iowa State University, has harnessed the power of WEPP to model erosion across Iowa in the Daily Erosion Project. The project is identifying locations prone to particularly high or low levels of erosion to enable interventions and understanding of the erosion process. Its utility is now broadly recognized, leading to its implementation beyond Iowa.[19]

All methods for estimating erosion are imperfect. Insufficient sampling can produce faulty conclusions with any of them. Some account only for soil movement from a location, others only its arrival to a new location. Geologists, for instance, tend to focus on measuring sediment in waterways, which is usually less than the amount of soil displaced from fields. The discrepancy between measurements of soil lost and sediments accumulated in waterways has led several geologists to suggest that the soil measurements overestimate erosion. But most critics do not account for the soil that is lost from agricultural fields and buried downslope, transported to a ditch, or diminished by transformation of organic matter to greenhouse gases. These all represent sources of soil that is eroded, is not available for agricultural production, and does not appear as sediment in waterways. Combining USLE with radioactive tracking and data from satellite images has helped to corroborate conclusions from each approach.[20]

Inaccessible and remote areas present an additional problem for generating accurate estimates of soil loss worldwide. Some erosion estimates in these regions have incorporated on-site information from farmers or researchers, but others rely entirely on remote sensing and GIS-based measurements. And the mountainous areas that are often hardest to reach are also where erosion is most severe.[21]

A big challenge to making sense of soil erosion is that it is reported as average rates across large land areas, but averages obscure local trends, which can be far higher or lower than the average. Erosion varies dramatically at different scales and across the globe. For example, although the global average of 13.5 tonnes per hectare per year might not immediately raise an alarm, contributing to that average is Fiji, a land losing soil at the fast clip of 50 tonnes per hectare annually. Likewise, annual erosion on cropland averages approximately 10 tonnes per hectare in the United States and 13 tonnes per hectare in the state of Iowa—rates similar to the global average. But in 2007, 2.4 million hectares in Iowa suffered losses of twice the state average. On May 6 of that year, 4 million hectares lost as much soil as is usually lost in an entire year, and 80,000 hectares lost 220 tonnes per hectare *in one rainstorm,* a loss more than one hundred times the soil renewal rate. After twenty such storms, a typical hectare of Iowa farmland endowed with 2,200 tonnes of soil will not have much remaining. In fact, about 4–17 percent of Iowa is in landscape positions most prone to erosion and for the most part devoid of topsoil, exposing geologic parent material (see pl. 6, top). So although the average for the United States or the state of Iowa might not be alarming, local loss may be more severe, rapidly stripping the soil and reducing productivity. In 2021 a shocking study revealed that across the Corn Belt of the United States about one-third of the agricultural land has already lost all of its topsoil.[22]

Since the beginning of agriculture, people have been accelerating erosion, whether we knew it or not. Thomas Jefferson, the third president of the United States, was a distinguished statesman, farmer, and architect. He was also a mass of contradictions. He wrote the Declaration of Independence, a brilliant tribute to human will and individual agency, yet he owned slaves throughout his life. He sought political counsel from Abigail Adams, wife of the second president of the United States, and yet he believed that women's only destiny was to serve men and raise children. He was also a scientific farmer, conducting extensive experiments to manage his 12,500-hectare plantation at Monticello, Virginia. Jefferson fervently believed in good land stewardship and yet spent five years inventing a new version of the moldboard plow, which has arguably caused more soil loss than any other implement in the history of agriculture (fig. 10). In 1813, Jefferson wrote in a letter that "the plough is to the farmer what the wand is to the sorcerer. It's [sic] effect is really like sorcery." And he pronounced deep plowing "a recipe for almost every good thing in farming."[23] As we'll see, he was colossally wrong!

Plows have been used since 3500 BCE to break ground for planting seeds. On uncultivated land, vegetation can be dense and difficult to penetrate with hand tools. The invention of the plow, which was originally made entirely of wood and pulled by animals, opened more of the Earth's surface to agriculture, thereby advancing the spread and productivity of agrarian societies. The moldboard was added to the plow to cut deeply into the soil, lifting and flipping it over 180 degrees. Jefferson's moldboard was made of iron, designed to be easier than its wooden predecessors for a horse to pull.

Figure 10. A modern recreation of Jefferson's "mouldboard plow of least resistance." Photograph printed with permission from © Thomas Jefferson Foundation at Monticello.

When agriculture moved west, farmers found that the iron plow didn't work as well as it did in the East. Designed for eastern soils, it rapidly became caked with the heavy midwestern soil, forcing farmers to stop every few meters to clean it. In 1837, a blacksmith named John Deere invented the first steel plow, which was heralded as a great advance because soil did not stick to the steel blade, and it was stronger than iron, enabling farmers to break ground that was previously thought unfarmable (fig. 11).[24] Deere built ten moldboard plows in 1839 and one hundred by 1842, creating the John Deere Company, a global farm implement business that still carries his name today.

Jefferson was partly right about the importance of the moldboard plow because it made the United States the agricultural powerhouse it is today. The capacity of the steel plow to break tough land expanded crop production across the Midwest and Great Plains in the nine-

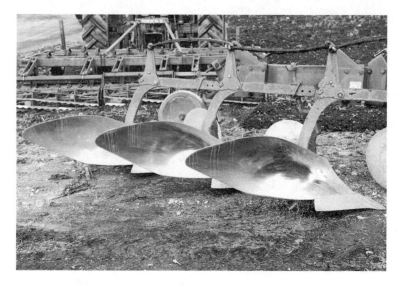

Figure 11. A modern moldboard plow. Photograph by Dwight Sipler.

teenth century. The agriculture that sprang from the newly opened earth also opened the rest of the United States to development, industrialization, and the many cultures that arrived with migrants exploring the western reaches of the continent.

The plow also had tragic consequences. It enabled droves of European settlers to move west where they displaced millions of indigenous people who had lived on the land for centuries. It also caused loss of much of the Midwest's soil and likely more than 25 percent of its carbon over the ensuing two hundred years.[25]

The plow became more than the tool of choice for breaking new ground. Its role grew to include turning cultivated soil every spring for planting, disrupting soil between crop rows to suppress weeds, and burying crop debris after harvest. Repeated tilling can cause erosion by directly moving soil downhill or off fields, but its greatest impact is the destruction of soil architecture, breaking clods into small particles vulnerable to movement by wind and water.

What prevents erosion? Plants are potent antidotes to soil movement. Hedgerows and wind breaks reduce wind velocity across cultivated fields. Trunks and stems hinder flow of rivulets, increasing the chance that water will penetrate the soil rather than run across its surface. Leaf canopy reduces the velocity of raindrops, which gently drip into the soil from the leaves that intercept them. Below ground, roots provide channels for the downward movement of water. Soil structure and water-holding capacity are improved by strands of glues produced by plants and bacteria that bind particles. Most plant species contribute to soil health, but the majestic giants of the world's forests are champions at soil protection. Their roots form immense subterranean networks that nourish the soil and anchor it to the terrain beneath.

Imagine, then, the impact of clear-cutting forests such as the tropical forests in the Amazon and Indonesia that are destroyed at a rate of approximately one football field per second, all day, every day. Societies throughout history have replaced forests with farmland to grow food and build shelters, but on sloped land the results can be calamitous if agricultural practices have exposed the soil to erosive forces. Several civilizations have collapsed or been forced to abandon their land after suffering widespread soil loss that resulted from deforestation.[26] Soil may be lost from denuded land gradually over centuries or rapidly over decades, depending on the quality of the initial soil, the slope of the land, the weather, and farming practices. Those agrarian societies that have persisted successfully on steep, formerly forested land are accomplished land stewards who have found ways to anchor the soil in place.

In the United States, deforestation began centuries ago. The stark consequences are evident in the Piedmont region that was clear-cut for farming. The Piedmont starts in New York State, spans Virginia and North Carolina, and extends into Georgia and Alabama. The region is not fit for agriculture without soil protection. Its steep and rolling hills are remnants of ancient mountain ranges of acid igneous rocks that spawned thin, sandy soils, carrying a mere 6–10 centimeters of topsoil. The forests had held the soil and fragile ecosystem in place, but from its settlement by Europeans in 1700 through the 1970s, the trees that were anchoring Piedmont soil were steadily replaced with fields. Plowing disrupted the soil, and crops such as tobacco sapped nutrients. Over time erosion on agricultural land increased one hundredfold compared with uncultivated areas of the Piedmont hills, eventually stripping most of the topsoil. Once the eastern Piedmont land became marginal, the settlers moved farther west to clear more forests for agriculture. Waves of migrants fled first to the middle Piedmont and then to its western edge in Georgia and Alabama in search of fertile soil. By 1967 Piedmont farming had largely ceased because the soil no longer supported it. By the end of the twentieth century the region had largely degenerated into scrub.[27]

The Piedmont may presage the outcome of erosion that is in progress across the United States and many other parts of the world. Europeans settled in the Midwest long after the Piedmont region, so the damage done by European Americans' agricultural methods appeared later. For example, in 1850 Minnesota's population was six thousand, whereas Virginia's population had surpassed that mark two hundred years earlier. Moreover, Midwest topsoil is deeper than Virginia's. So it is not surprising that the kind of wasteland soil evident in the Piedmont is rare in the Midwest. But Minnesota's erosion is estimated to have increased one hundredfold since arrival of European settlers, suggesting that its trajectory may be similar to the Piedmont's, just delayed. With estimates of global soil loss due to agriculture between 75 gigatonnes per year and more than 130 gigatonnes per year (exceeding the rate of soil genesis by thirty-seven to sixty-five times), it is likely that many other regions are following the pattern of the Piedmont.[28] Another pattern that is recapitulated across the globe is that exhausted soil leads to the abandonment of land, followed by human migration.

The Piedmont's agricultural history also shows us that the transition from perennial to annual plants can ravage the soil. In this case, planting annual row crops such as cotton and tobacco sacrificed the ancient, sprawling root systems of the perennial trees they replaced. The life strategies of annuals and perennials differ in that annuals complete their lives in a single season, and perennials go dormant in the winter and then grow again in the spring. Annuals reproduce only through seed, whereas perennials' resilient roots persist from year to year, giving them renewed life each spring in one location while their seeds spread them to others. To maximize the potential of each life strategy, at the end of the growing season annuals dedicate their photosynthetic resources to creating seeds that will serve as their genetic legacy, whereas perennials invest energy in their roots—the organ that gives them life in the next season. Deforestation is just one practice that forfeits the underground influx of gifts from perennial plants that give soil its robust structure and nourish its inhabitants.

Like deforestation, the conversion of prairies to agricultural crops has replaced perennial with annual plants. Not surprisingly, this is another kind of land use that has accelerated erosion. The vast expanses of prairies and steppes that once covered almost 2.5 billion hectares of the world's fertile, black Mollisols were home to hundreds of perennial plant species, which annually flaunt their diversity above ground as an indicator of the industry going on below.[29] It is the root systems of these plants that make possible the spring-through-autumn pageant of flowers: the pink sweep of prairie shooting stars followed by expanses of majestic blue lupines, the sizzling orange Indian paintbrush, the charming yellow cup flower, and hundreds of others. The delicate tassels of perennial grasses dance in summer breezes before their foliage turns shades of brown and purple as the plants begin to prepare their roots for winter. The aboveground prairie beauty is a colorful celebration of the powerful root systems that hoard nourishment to survive harsh winters and return to active duty in the spring.

The roots of perennial grasses and legumes are often larger than their stems, leaves, and flowers—an underground cache that expands each year. The roots of perennial grasses such as switchgrass, for example, contain 50 percent of the plant's biomass during its first year of growth, and by the third year the roots have outstripped the shoots to garner 80 percent of the plant's biomass, extending more than 4 meters down the soil profile. Perennial roots also turn over rapidly, with 30–86 percent replaced each year.[30] The decomposition of this prodigious underground filigree forms deep soils containing copious organic matter that have provided abundant harvests around the world. Now these prairie soils are under threat.

Because plants have a finite amount of carbon to distribute to their organs, abundant seed production is often associated with diminutive root systems (fig. 12). When people started breeding annual crop plants to optimize seed production, their roots shrank further. Today the roots of a typical corn or wheat plant represent only 40 percent of the plant's biomass during the growing season and as little as 3 percent by harvest, leaving a trifling amount of carbon to restock the soil. In the United States 99 percent of the original prairies are now used for agriculture, including much of the nation's 225 million hectares of corn and 157 million of wheat.[31] It was this conversion that led to the Dust Bowl.

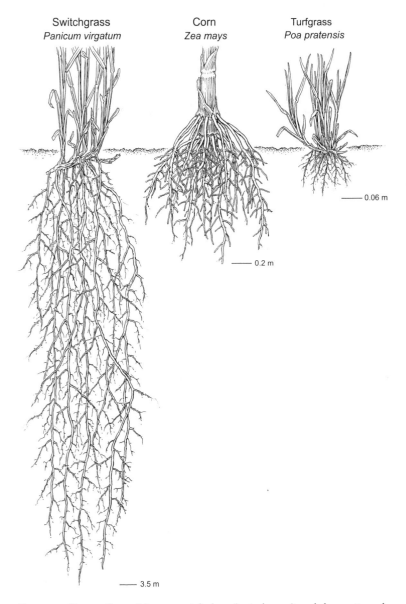

Switchgrass
Panicum virgatum

Corn
Zea mays

Turfgrass
Poa pratensis

—— 0.06 m

—— 0.2 m

—— 3.5 m

Figure 12. Roots of a prairie perennial plant (switchgrass) and domesticated corn and turfgrass. Illustration by Bobbi Angell.

The decline of the Piedmont region and the devastation of Oklahoma and Kansas during the Dust Bowl era bring into sharp relief contrasting causes of soil erosion. The Piedmont started with thin, forested soils that were diminished by deforestation and the cultivation of dense row crops. The steep slopes heightened the land's vulnerability to water erosion, and without trees, the fragile topsoil was rapidly washed away. In contrast, the plains of Oklahoma and Kansas are flat, but the transition from perennial prairie plants to annual crops made the soil susceptible to forceful winds during the drought of the 1930s. These erosion tragedies tell the same story. No matter whether topography or weather was the inciting factor, the decline of the soil in both instances can be attributed to the conversion of wild perennials with undisturbed soil to annual crops with meager root systems and ravaging deep soil plowing.

Given that the Virginia portion of the Piedmont area had already suffered extensive soil degradation by the time Jefferson invented his iron plow, it is surprising that a scientist with his commitment to land stewardship did not realize the impact of plowing on soil loss. In the same letter in which he extolled the virtues of the plow, he advocated reducing erosion by planting across hillsides rather than up and down them, a soil-protective practice known today as contour farming. But he stubbornly continued to believe that his plow was an entirely good thing for the land and even blamed "evil" rain for carrying soil off the land.[32] Whether or not Jefferson was aware of the connection, now it is well established that frequent land cultivation, particularly with a moldboard plow, hastens erosion. Today Jefferson's legacy is deeply tarnished by his slave-holding throughout his life. Less broadly appreciated is how his iron plow enabled European Americans to farm and populate the Midwest, thereby contributing— along with political, economic, and military objectives—to the devastation of many Native American populations and the destruction of the region's soil structure, paving the way for the Dust Bowl one hundred years later.

Cultivating crops is not the only way that agriculture erodes soil. Cattle and other hooved farm animals can completely transform the landscapes they traverse, degrading soil in several ways. When they are allowed to overgraze, livestock consume foliage down to soil level, dislodge plants, and prevent regrowth. Heavy animal traffic compacts the soil so that water cannot easily penetrate. Over time, plant growth is diminished, and the soil becomes drier and more erodible. Water scarcity, dwindling vegetation, and erosion become locked in a negative feedback loop that causes a spiraling ecosystem decline.[33]

People have been constructing edifices for a long time, but buildings began to have noticeable effects on soil in the twentieth century, especially as modern cities expanded. The drive toward urbanization is reducing the amount of land available for agriculture at a rate of 1.6–3.3 million hectares per year—that's the equivalent of the area of Lebanon and Belgium, respectively.[34] The choice of construction materials is also consequential. Impenetrable concrete and asphalt, for example, leave water without a route into the ground, which leads to flooding and soil erosion.

Of all types of construction, none has such visible impacts on soil as damming rivers. Rivers and dams exemplify the paradox of erosion as a benefit and a curse. Although upstream erosion may deplete the land, rivers often deposit silt on their banks and at the coast where they spill into the sea, thereby creating fertile floodplains and preventing shoreline erosion. But today, these processes are threatened by thousands of dams across the world.

Between 1960 and 1970, the Aswan High Dam was built to regulate the flow of the longest river in the world.[35] The majestic Nile traces a south-to-north trajectory through more than half of the African continent. Originating in Burundi, the White Nile River snakes north through Uganda, South Sudan, and Sudan. There it meets the Blue Nile River, sourced in Ethiopia. The two rivers merge to become the Nile River, which continues through Egypt, concluding its journey on Africa's northeastern coast, where it opens its massive mouth and flows into the Mediterranean Sea. Water from the Burundi source will take three months to reach the sea, traversing a 6,695-kilometer path, at times flowing gently, at others roaring along at 3 meters per second. The Nile erodes lands as it flows, collecting silt that it redistributes to the Nile Delta—the 20,000-square-kilometer region flanking the river—during floods, and to the coastline as it empties into the Mediterranean.

The purpose of the Aswan High Dam was twofold: to prevent the floods and droughts that plagued farmers in the Nile Delta and to provide hydroelectric power to the Egyptian people. Designed by a group of British engineers and built by a Russian team, the Aswan High Dam is a towering testament to human ingenuity. The dam—a rock and clay wall—stands 111 meters high and 3,830 meters across. It impounds 169 billion cubic meters of water that form a reservoir, Lake Nasser, which collects above the dam and extends 320 kilometers upstream in Egypt and another 160 kilometers in Sudan.

The Aswan High Dam accomplishes its intended tasks proficiently, discharging water at a measured cadence and generating 10 billion kilowatt-hours of electricity annually, enough to serve almost half of Egypt's population. But it also caused an unintended consequence. When the river reaches the Aswan High Dam, the flowing water comes to a halt, pooling in Lake Nasser. As the water is waiting to be released, it is stilled. Without the river's turbulence keeping silt particles suspended, they settle to the bottom of the lake. As a result,

98 percent of the silt does not traverse the dam, remaining in the reservoir and not deposited along the rest of the river. The Nile once made an annual delivery of 10 million tonnes of sediment to the surrounding delta on its way to the sea and 124 million tonnes when it reached its destination, the Mediterranean Sea. Today that sediment never reaches the riverbanks and ocean shorelines, leaving them without reinforcements to combat their own erosion. The downstream sediment deficit is causing the Nile's banks to recede, some at a rate of 125–175 meters per year. Likewise, the Mediterranean shoreline where the Nile empties is retreating rapidly.[36]

The dam has also starved the Nile Delta, source of two-thirds of Egypt's food production. The farmland now requires chemical fertilizers to replace the 7,000–10,000 tonnes of phosphorus, 7,000 tonnes of nitrogen, and 110,000 tonnes of silica once contributed by the Nile floods.[37] Some experts argue that the Nile no longer has a true delta because of the deficit of silt.

The Nile is not alone in its dam plight. Most of the world's great rivers have been dammed—the Amazon River, Yellow River, Columbia River, Colorado River, the River Tigris—either to regulate water flow or produce hydroelectric power. Each dam has had complex impacts on the land and people around it. Across the world, human intervention has increased sediment flowing through rivers by 2 gigatonnes and simultaneously reduced the amount reaching the coast by 100 gigatonnes because most of the sediment is trapped in dams. Today, silt fills one-fifth of the storage capacity, or 1,100 cubic kilometers, of the world's reservoirs above the dams and costs $2–$3 billion worldwide for repairs to damaged turbines and loss of power generation in hydroelectric plants.[38] Some have caused landslides, many have permanently altered agriculture downstream, and all have created both challenges and opportunities for local wildlife.

From the first dam in human history (the Jawa Dam in Jordan, built around 3000 BCE) to the first hydroelectric dam (built in Appleton, Wisconsin, in 1882) to the largest dam in history (the Three Gorges Dam on the Yangtze River in China, completed in 2006), each served a critical function for people, and all have altered the surrounding landscape. As with most technology, society needs to weigh the benefits of these wonders of engineering against their alterations of the natural world. The soil should receive consideration in this calculus.

Climate is a hybrid of natural and unnatural forces that drive soil erosion. Although climate events accompanied by high-velocity wind and water have always driven erosion, today the climate has taken a sharp, unnatural turn amplified by human activity. If climate is a natural driver of erosion, then anthropogenic climate change is an unnatural manifestation that is devastating soil. The effects vary by region. Some already suffer severe soil loss, and others will feel the effects in the near future. All indicators point to an increase in water erosion resulting from severe rainstorms that will become more common around the world. The more severe the storm, the worse for soil, because high-velocity raindrops carry the force needed to detach and move particles. The frequency of severe storms escalated in Asia, Europe, the Northern Territory of Australia, and North America between 1964 and 2014. This trend is strikingly apparent in U.S. records: for the first half of the twentieth century, the annual number of heavy precipitation events clustered around a mean, but since the 1950s, the frequency has steadily, unrelentingly risen (fig. 13).[39] The timing of these storms determines their impact. Early and late in the season, if the soil has been plowed but not yet planted, there are few barriers to soil movement and nothing to impede the raindrops. In the middle of the season, when crops are thirstiest and protect the soil, severe rain can be less damaging. But if the storm produces hail,

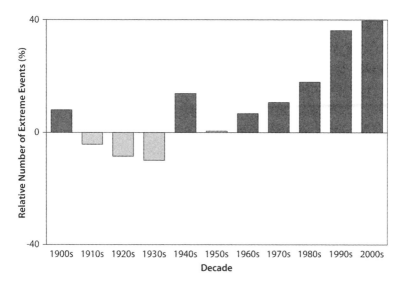

Figure 13. Heavy precipitation events in the United States since 1900. Illustration by Bill Nelson adapted from J. D. Walsh et al., "Our Changing Climate," in *Climate Change Impacts in the United States: The Third National Climate Assessment,* ed. Jerry M. Melillo, Terese Richmond, and Gary W. Yohe (Washington, D.C.: U.S. Global Change Research Program, 2014), 19–67.

an entire crop can be lost in a single devastating event. Imagine the damage done by hailstones measuring 22 centimeters in diameter, like the one in Argentina in 2018 that broke the world record! Severe storms are likely to intensify with continued climate disruption, and water (in both liquid and solid forms) will regularly pummel the land's surface, challenging the soil and its stewards.[40]

Climate change is typified by extremes of wet and dry, wind and heat. While some parts of the world are being bombarded with water, other parts are becoming parched, contributing to another form of land degradation—desertification. The world population experiencing desertification since 1961 has doubled, and the frequency of dust storms has increased sharply. The interaction of hotter temperatures, a more variable climate with less rainfall in some regions, and changing ways of using land—including urbanization—have driven a surge in desertification, which in turn promotes soil erosion. Droughts have increased in Amazonia, northeastern Brazil, the Mediterranean, Patagonia, most of Africa, and northeastern China.[41] As global warming marches ahead unchecked, soil erosion and degradation are bound to increase. People drive soil erosion indirectly by accelerating climate change. We drive it directly by how we treat the land.

As farmers continue to plow, as cities continue to sprawl, and climate change continues to intensify the very weather patterns that cause erosion, more topsoil will disappear. During the twentieth century, scientists debated whether soil was not, in fact, a renewable resource. We no longer have the luxury of continuing this debate in the face of soil loss rates that exceed agricultural rates by one hundredfold in some regions. In 2015, the International Year of Soils, the United Nations declared soil finite and predicted a catastrophic loss within sixty years.[42]

But why does extinction of topsoil constitute a crisis? And what are the effects along the way, when only some of the soil has been lost? To answer, we must explore the impacts of erosion around the planet.

6

Rocky Planet

Imagine standing on a planet with a rocky surface that emits no odor and is unable to sustain life. A gust of wind stirs sand particles into the air, obscuring a bright blue sky. When it rains, rivulets carry sand and gravel to gullies and then rivers, filling them with sediment. Silt, pebbles, and boulders have replaced the spongy, fragrant carpet that once gave the land its life force. This is a world without soil.

There will always be soil on Earth as long as there are forests and prairies to nurture it, but what if much of our sloped agricultural land lost its fertile topsoil and began to look as if it belonged on a rocky planet? The loss of food security would be enormous, the loss of beauty incalculable. We are not at the rocky planet stage now, but erosion is already affecting landscapes and food production worldwide (see pl. 2). As soil loss advances, the threat to global food security will increase and safety nets will shrink, generating unprecedented hunger in parts of the world that never before experienced food insufficiency as well as in those places all too familiar with it.

To collaboratively tackle urgent global issues such as soil erosion, we must first understand the complex connections that unite us through food and soil. The food system is complex because part of it reverberates far beyond its visible reaches. Every country—indeed, every person—is connected to the rest of the Earth and its inhabitants because we participate in global markets, share food, and breathe the same atmosphere. And yet every farm is unique, each wrestling with a different constellation of variables. The physical environment, the biology of plants and animals, and the social forces that drive human behavior intersect at the farm level and can be difficult to disentangle. The far-reaching consequences of soil erosion lead to an exploration of the intersection of soil with crop cultivation, biodiversity, the role of women in agriculture, the impact of change on small landholders, dams and hydroelectric power, and the qualities of each society that influence its policies pertaining to soil. The similar effects of soil-related issues on various countries reveal universal themes, and the differences illuminate singular features of each country's relationship with soil. Exploring the consequences of soil erosion in diverse countries under changing conditions is, therefore, not solely a scientific endeavor.

How should the impacts of soil erosion be measured? As lost income? Reduced crop yields? Lost biodiversity? Or in centimeters of a nonrenewable resource? The United Nations reports that cumulative land degradation—80 percent of it caused by soil erosion—is harming the welfare of 40 percent of the world's population, contributing to global and regional conflict, and causing mass migrations. An economic model that includes direct and indirect impacts of soil erosion predicts $23 trillion in losses of food, ecosystem services, and income worldwide by 2037, with sub-Saharan Africa losing the largest share, at 16 percent. Today, crop production, biodiversity, and ecosystem services lost with eroding soil amount to 10 percent of annual global gross product.[1] Some of the losses are irrevocable.

Without soil, agriculture grinds to a halt. Long before the soil is entirely gone, erosion cuts into crop production. Already, erosion reduces crop yields by an estimated 0.3 percent each year, which will likely lead to a cumulative loss of more than 10 percent of crop production worldwide between now and 2050, when the Earth will be groaning under the weight of feeding nine billion people.[2] Global averages mask the local impacts where soil loss becomes personal, devastating farmers and their families. The worldwide average erosion of 13.5 tonnes per hectare each year is meaningless to the farmers whose slopes are washing away at a rate of 100 tonnes per hectare per year, robbing them of the only means they have to provide for their families. Farmers who dream of passing on a robust farm to their children may be far more troubled by erosion than companies that rent cropland temporarily.

The imminence of erosion's impacts on food production varies with soil type. The global average erosion rate manifests differently in various locations, varying with soil age, texture, and depth. But confronted with losses ten to one hundred times greater than the rate of soil production, even the deepest soils cannot sustain agricultural productivity for long. Imagine a fertile Mollisol endowed with 2,200 tonnes of topsoil per hectare that is losing soil at the world's average erosion rate of 13.5 tonnes each year and producing it at one-fortieth of that pace. Crop productivity would be affected within decades, and

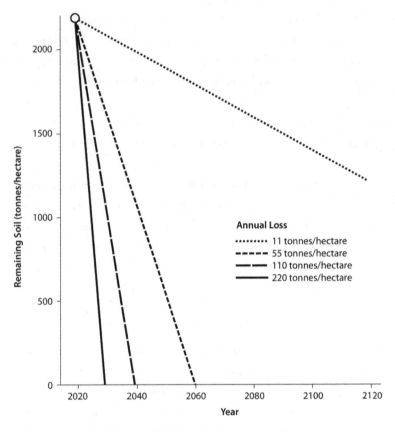

Figure 14. Projected soil erosion from land starting with 2,200 tonnes per hectare. Illustration by Bill Nelson and Marc G. Chevrette.

the topsoil—the fertile O and A horizons—would run out in about two hundred years. If erosion accelerated to 55 tonnes per hectare annually, the land would be devoid of topsoil in forty years. At 220 tonnes per hectare, this Mollisol would go extinct within a decade (fig. 14). The deep Mollisol poses a best case because most of Earth's land has far less than 2,200 tonnes of soil per hectare, but even the fertile Mollisols of the Midwest are suffering: one-third of the crop-land has already lost the entire topsoil.[3]

In Africa, a largely agricultural continent, erosion poses a severe threat to food production. Compared with the highly productive Mollisol, which might have 20–150 centimeters of topsoil, soils in Africa are typically less fertile and more shallow, many sporting less than 10 centimeters of topsoil. Ultisols of sub-Saharan Nigeria are often highly degraded, with only a thin top layer remaining. For individual farmers and their families, losing 50 tonnes per hectare each year would be catastrophic. This scenario is in sight. Anambra, a state in southeastern Nigeria, is scarred by barren gullies that form following expansion and contraction of land during wet and dry cycles and provide a route for rapid soil migration. In areas with minimal vegetation, gullies remodel the landscape, growing so massive that one researcher suggested renaming them canyons. Indeed, one gully measured almost 3 kilometers long by 349 meters wide. It is no surprise, then, that annual erosion in some locations in Anambra can reach as high as 2,200 tonnes per hectare! In the neighboring state of Imo, erosion rates range from 6 to 1,200 tonnes per hectare per year, with an annual mean loss of 36 tonnes per hectare.[4] About 20 percent of the land erodes at a rate of at least 235 tonnes per hectare each year—the consequence of high rainfall, slopes, and sparse vegetation. That means that farms on one-fifth of Imo's land could have little topsoil left and crop yields will plummet within a decade if erosion continues unchecked.

Farm productivity is influenced by several interacting factors, making it difficult to isolate the impact of erosion on crop yields. To establish this connection, one group of researchers conducted a "de-surfacing" experiment in which they removed the top layers of soil at three sites in Nigeria, keeping all other conditions the same, and then measured crop yields in the remaining soil. Desurfacing treatments at all three sites reduced yields, but most dramatically at Onne, where maize yield was reduced by 95 percent, 95 percent, and 100 percent when 5, 10, and 20 centimeters of soil, respectively, were removed.[5] The study is a stinging reminder that soil loss reduces crop yields, a problematic conclusion for the food supply of a country where land can lose 2,200 tonnes of soil per hectare in a single year. At these rates of erosion, the inability to raise crops is only centimeters away. Once the soil is gone, it is difficult to restore productivity, leading farmers to abandon many degraded soils.

In southern Nigeria, farmers report watching floods wash away their cassava, yam, and cocoa crops and, with them, the soil. Local food shortages worsen as rivers fill with sand, preventing fishermen from catching small fish to use as bait for larger ones. In northern Nigeria, where the climate is characteristically hot and dry, crops grown in unhealthy or eroded soil are more susceptible to drought. As erosion transforms the landscape, its impact creeps into people's lives, threatening their livelihoods and local economy. In fact, soil erosion is likely one of the factors that placed northern Nigeria on the list of eight countries experiencing the worst food crises in 2018, when an estimated 28 percent of Nigerians suffered from hunger.[6] Past food insecurity, a growing population, and violent conflict with the Boko Haram bode poorly for Nigeria's ability to meet its food needs in the future. Continued soil loss reduces the yield potential, capping the amount of food that could be produced under the best conditions and leading to inevitable crop loss under the worst.

Morocco sits on the northwest coast of Africa bordered by the Mediterranean Sea. Its culture is distinguished by diverse influences from Europe, Arabia, and the ancient Berber tribes that have inhabited the region over the last twelve millennia. Morocco's agriculture is similarly diverse, including cereals, fruits, nuts, and livestock that buffer the nation's diet and economy against dependence on any one product. Despite its stable agricultural architecture, Morocco faces land devastation. Almost 70 percent of Morocco's land area is agricultural, and about half is degraded by soil erosion, with mountainous regions losing 50–400 tonnes of soil per hectare annually. Climate change has intensified Morocco's droughts that make small farms vulnerable to crop loss and being swallowed up by large conglomerates that grow cash crops for export. The share of agriculture in Morocco's gross domestic product (GDP) has stagnated at 12.4 percent, largely as a consequence of low crop yields, despite government efforts to promote agriculture. But Morocco has an edge in global markets because of its location, which poises it to be a strategic hub for corporations wanting to operate in Africa.[7] The success of its global alliances depends on continued food production and, thus, reduced erosion. Given the country's vulnerability to drought, which is likely to intensify as a consequence of climate change, the Food and Agriculture Organization has selected Morocco for a pilot project to test the effectiveness of its new Sustainable Food and Agriculture program that aims to stabilize the food supply.

Erosion poses an even greater threat to countries on the eastern coast of Africa. Ethiopia, for instance, loses 1 billion tonnes of topsoil annually, causing moderate to severe land degradation on 25 percent of the land and costing the nation 3 percent of its agricultural GDP. Farther south in East Africa lies Tanzania, where land degradation has ranked as the top environmental problem for more than sixty years, largely because erosion affects 61 percent of the land. Tucked between Ethiopia and Tanzania, Kenya is another country with complex soil challenges caused by deforestation, overgrazing, and agricultural practices. Some political leaders have expressed the hope that cell phones, now owned by 70 percent of the population, will provide ready access to advice on farming methods and adoption of sustainable soil practices, but the problem is due more to poverty than to lack of information. Kenyan women, who grow most of the country's food, have extensive knowledge of soil and how to manage it well, but they often lack time or financial resources to implement best practices. Many juggle land management with several other jobs that are needed to sustain their families, and they choose unsustainable farming methods in order to devote more time and money to their families, a choice unlikely to be altered by cell phones. Farmers throughout the developing world face similar trade-offs, often choosing to confront more immediate problems than erosion. The exigency of survival is the overwhelming concern for almost two-thirds of the world's poorest adults who work in agriculture, an uncertain endeavor plagued by the vagaries of nature and economic markets, creating a constant state of financial crisis and risk.[8]

In Asia, the largest continent, effects of erosion are as diverse as the region's terrains and climates. The impacts span deserts and rain forests, coastal flatlands and the world's tallest mountains. Soils vary from youthful Entisols to permanently frozen Gelisols to soggy Histosols, to name but a few. Asia's crop species also vary by region. India and China lead the world in rice production. Central Asia abounds with wheat, cotton, and sugar beets, and Southeast Asia produces maize, coffee, cocoa, tea, coconuts, and rubber. The continent is also known for its colorful cornucopia of fruits—bananas, pineapples, citrus, papayas, durian, lychee, and mangosteens—grown for fresh consumption, export, and canning. The factors driving soil loss are similarly diverse, but common to all instances of erosion is the toll it takes on farmers, the food supply, and the broader economy.

Almost half of South Asia's agricultural land is degraded, causing $10 billion in annual losses, or 2 percent of combined GDP and 7 percent of agricultural output. Once again, averages fail to capture the local impact because land degradation is uneven, varying with soil type, topography, and farming practices. The extent of land degradation varies across Bangladesh (65–75 percent of land area), Pakistan (39–61 percent), and Bhutan (3–10 percent), as do its consequences. In Bangladesh, 75 percent of the hilly areas are highly susceptible to water erosion, which costs $140 million in losses to cereal production and $544 million in overall nutrient loss. These already considerable effects on food production are aggravated by another trend: land use conversion. Each year, Bangladesh converts up to 80,000 hectares of farmland to nonagricultural uses, such as construction of houses, roads, markets, schools, and industry, resulting in an annual reduction of 1.6 million tonnes of domestically produced food.[9] Although Bangladesh has undergone sustained economic growth since the 1980s and poverty has declined, the World Food Programme reports that a quarter of all Bangladeshis have insufficient food and eleven million live with acute hunger. In light of current hunger and the horrifying memory of Bangladesh's famine in 1974, further food losses are unacceptable.

Bhutan serves as a solemn reminder of the need for fine-grained analysis of national statistics. The serene landscape of this tiny green kingdom is nestled between China and India on the southern slopes of the Himalayas. Often described as the most beautiful country in the world, Bhutan is known for its picturesque mountains and gushing rivers that deliver Himalayan ice melt to lower altitudes. The country is also famed for its environmental policies, which make it the only carbon-negative country in the world—that means that more carbon is fixed by Bhutan's vegetation than the country releases into the atmosphere. Its singular identity is a consequence of Bhutan's verdant forests, which are constitutionally mandated to cover at least 60 percent of the land.[10]

At first glance, degradation of "only" 3–10 percent of the land might seem like good news for Bhutanese farmers. Not so. Agricultural land is scarce, producing terraced rice paddies neatly stacked on mountainsides and small plots of maize cultivated by 37,000 households largely at a subsistence level. Although agriculture represents only 22 percent of the GDP, 69 percent of the population relies on land to make a living. On average, each family owns 1.4 hectares of land, and 60 percent of households have less than that. Since soil erosion is concentrated on agricultural land and most of the country is forested, the national average substantially underestimates the impact of soil erosion on impoverished farms clinging to the sides of mountains. Fortunately, the government has prioritized soil protection in its environmental policies.[11]

The royal Bhutanese government is philosophically committed to protecting its citizens and its majestic landscape that hosts extraordinary biodiversity, including 105 plant species not found anyplace else. Bhutan is the only country to adhere to a different measure— Gross National Happiness—rather than GDP to create and assess national policy. The constitution is threaded with Mahayana Buddhist values that emphasize the relationship between human well-being and coexistence with the natural world. Accordingly, policies are designed to both feed the human population and protect Bhutan's tremendous natural resources. In addition to protecting soil for food production and for its own value, Bhutan's government has made erosion a national priority because of its impact on the hydroelectric industry, the country's largest source of income. Soil erosion dumps sediments into power plants, damaging turbines and generating 60 percent of total repair costs. So far, policies to curb erosion have encouraged Bhutanese farmers to terrace steep slopes for rice cultivation and forbade clearing land and leaving it bare for several years.[12] Although erosion has been extensive in Bhutan, this little kingdom is committed to the task of protecting and rebuilding its soil.

Moving closer to the equator in South Asia, the island of Java stands out for the conflict between its fragile, mountainous terrain and high demands on the state's agriculture. Java produces half of the agricultural products of archipelagic Indonesia, a republic of 273 million people and the world's fourth most populous country. The mountains that cover half of Java do not deter intrepid farmers from raising crops on these volcanic highlands. On the steepest land maize and cassava are preferred for their low maintenance, especially when the farms are far from the farmers' villages. Rice is also common despite its labor-intensive cultivation. In central Java, flat land erodes at around 25 tonnes per hectare each year, whereas steep land surpasses 200 tonnes of soil per hectare annually, with agricultural land suffering the worst erosion, topping out above 300 tonnes per hectare annually. The resulting yield losses shrink the total GDP and the agricultural economy. This is not the right trend for a country that has seen little growth in food production since the late twentieth century and has experienced a concurrent 30 percent population increase. Sediment records indicate that erosion tracks closely with population density, accelerating sixfold during the twentieth century as the Javanese population swelled from twenty-eight million to well over one hundred million.[13] Although a causal relation between population growth and soil erosion has not been established, the pressure to ramp up food production often encourages the use of soil-degrading practices. Indonesia's projected population growth of 22 percent by 2050 will surely stress its entire agricultural system. The poverty afflicting many small producers is likely to worsen as the population pressure intensifies, which in turn will increase the use of farming practices that damage soil. Particularly on small farms, soil is a likely casualty of the projected population expansion in Indonesia.

Smallholders, people who produce or gather food on a small scale, are the most vulnerable to the effects of soil erosion. Smallholders represent 93 percent of Indonesian farmers and grow the bulk of such staple crops as rice, corn, and cassava as well as the cash crops of coffee, tea, spices, fruits, and vegetables. Remarkably, this tremendous spectrum of crops is grown on farms that average 0.6 hectares. Worldwide, there are 1.5 billion smallholders making a living from less than 10 hectares of land. They work as farmers, pastoralists, and forest keepers and manage 80 percent of the farmland in Asia and sub-Saharan Africa. Smallholders are among the poorest farmers worldwide, with modest plots of land supporting families often living below the poverty line. They are often the most vulnerable to natural disasters that destroy crops; armed conflicts that cut off food supplies; and the vagaries of world markets buffeted by unpredictable fluctuations in supply, demand, and commodity prices.[14]

Women constitute 43 percent of the global agricultural labor force. In sub-Saharan Africa, where women produce 80 percent of the food, traditions and laws often prevent them from owning land, and their only route to farming is through men. Women also typically have fewer opportunities for off-farm employment, making them more vulnerable to economic hardships caused by weather, soil exhaustion, and war.[15] As food becomes scarcer, women will bear a disproportionate share of the burden. The impact of erosion on food production will touch everyone but will leave some communities in ruins and many women's livelihoods in jeopardy.

The preponderance of small farms worldwide means that the well-being of smallholders is critical to global food stability and land stewardship. Smallholders are the guardians of precious genetic diversity in crop plants. Only twelve plant and five animal species produce 75 percent of the world's food, and large-scale agricultural systems often select highly inbred, uniform members of these species, a process that narrows the gene pool. In contrast, smallholders tend to raise food from less highly bred local lines that retain some of the genetic diversity of the wild relatives from which they evolved. As more challenges confront the agricultural system, these genetic reservoirs are likely to be the source of traits coveted by breeders worldwide, including drought tolerance and resistance to disease, which will fortify crop and livestock production as the climate changes.[16] When smallholders end their agricultural efforts, their genetic resources can be lost because the valuable genes in their plants may not be found in any other members of the species.

Spinning the globe to the west, we also find land degradation eroding economies. Soil in Ukraine is worthy of reflection for its sheer size, as the second largest country in Europe, and for the paradox of finding the world's richest soil in one of Europe's poorest countries. Two-thirds of Ukraine, or about 28 million hectares, is thickly blanketed with one-third of the world's blackest soil, Chernozem, which is Russian for black soil (see pl. 5 top).[17] This Mollisol is special, with an organic horizon that can be as deep as 1.5 meters, creating some of Earth's most fertile land, supporting legendary agricultural productivity, and earning Ukraine its moniker, "the breadbasket of Europe." Spanning the width of the country, the Chernozem region generates Europe's highest production of grain and potatoes and is the world's premier producer of sunflower seeds. In the spring, bright green shoots stand out against the ebony land, followed by sweeps of dense green corn, the blue haze of barley, and brilliant yellow sunflowers stretching endlessly toward the horizon. In autumn, the tans and browns of the harvest are converted to gold, as farmers sell grain and other staples locally and through the world market. Its prime geographic location makes Ukraine a key food exporter to the Russian Federation and the European Union, and its deep ports along the Black Sea provide access to Middle Eastern and North African markets.

To those who revere soil, Ukraine is a venerated destination. Few have ever seen land with 150 centimeters of jet-black topsoil, its richness exhibited in the extraordinary color and the dark fragrance of geosmin. For a long time, there was a pervasive belief that the great Chernozem was invulnerable and somehow resilient to any insult thrown its way. This misconception delayed action to protect the soil. The Ukraine Chernozem (see pl. 5, top) started with a lot more soil than most places in the world. It also has more to lose.[18] And losing it is. At rates that are cause for protest (see pl. 5, bottom).

For every tonne of grain produced, Ukraine loses 10 tonnes of Chernozem, and for every agricultural dollar generated, one-third is lost in the eroding value of the soil. The country's agricultural land is stripped of a combined 500 million tonnes of black soil each year, which has begun to reduce crop yields by up to 50 percent (see pl. 5 bottom). The peril of losing the great Chernozem and with it a mammoth food producer has ignited concern in the global community, which united around the Ukrainian Soil Partnership, an organization launched in 2019, whose goal is to preserve soil and eliminate Ukraine's land degradation by 2030. The partnership offers training to farmers and scientists and is seeding demonstration plots to illustrate best soil conservation practices.[19] If this effort is unsuccessful, how long will the Chernozem last, and how long can Ukraine remain Europe's agricultural powerhouse?

Ukraine is not the only country endowed with Mollisols that are washing away along with food and income. The United States, which is home to 200 million hectares of Mollisols, has been losing soil since European colonists arrived. Researchers have tried to estimate the cost of soil erosion and its impact on agriculture for decades but have found it difficult to isolate erosion's impact from all the other factors influencing yield and profit. In 1933, the cost was estimated at $3 billion annually, which is today's equivalent of $58 billion, and sixty years later it was approximated at $44 billion for combined costs of damages and preventative measures.[20]

The costs of soil accrue from several directions. On the eroded land, nutrient loss, yield reduction, land devaluation, and loss of biodiversity must all be accounted for in cost estimates. Off-site erosion costs range from sediment deposition, flooding, need for water treatment, increased food prices, and intensifying climate change.[21]

Some of the worst erosion in the United States is in the Midwest, where our valuable Mollisols are concentrated. It was during a phone call from the White House about water erosion in Iowa with soil scientist Rick Cruse that I began to realize that soil erosion is in fact a slow-burning national crisis for the United States. The state of Iowa illustrates erosion's impact in the United States. Situated on land that was midwestern prairie before it was farmed, Iowa's Mollisols support intense cropping systems and high yields. Iowa generates the second highest agricultural revenue among the fifty U.S. states for its production of corn, soybeans, and pork. As an exporter of about $11 billion of agricultural commodities annually, Iowa is a prominent player on the world stage.[22] If it were a country, Iowa would be the fourth largest producer of corn and soybeans worldwide!

Despite Iowa's relatively gentle slopes, the state has lost enough soil to see disturbing yield reductions, and the projections for the future are bleak. In 2007, Iowa reported that 4 million hectares had each lost 11 tonnes of topsoil and an additional 2.4 million hectares had eroded at twice that rate. The amount of land eroding in Iowa and Ukraine is about equal, but Iowa's topsoil is less than one-tenth the depth of Ukraine's and is much closer to hitting subsoil. In fact, bare parent material, exposed by erosion of topsoil, is visible in many locations across Iowa (see pl. 6, top). Overall, Iowa's annual erosion rate is roughly equivalent to the world average and ten times the rate of soil genesis. More disturbing are the rates of 55 tonnes of soil per hectare annual loss from 200,000 hectares and 220 tonnes per hectare on 55,000 hectares. Without intervention, this land would be devoid of topsoil in forty years and some much sooner. Even before bottoming out, crop yields decline as the soil loses fertility and depth. The financial costs of erosion in Iowa are growing, predicted to reach $315 million in ten years and $735 million after fifteen years. Depressed yields will be as unevenly distributed as erosion itself; the financial security of farms losing 50 tonnes per hectare annually will suffer more than those experiencing negligible erosion. Farming is a precarious business in much of the world, and Iowa is no exception. With profit margins of 4–13 percent in Iowa agriculture, even small yield losses can undermine a farm's financial stability; large losses are untenable. In many regions, including the Midwest, soil depth appears to drive yield. A study in Ohio reported that removing the top 20 centimeters of soil reduced maize yield by 50 percent. The dependence of yield on soil makes for a gloomy outlook for Iowa's farms experiencing erosion (see pl. 6 bottom).[23]

Unfortunately, the rate of erosion in Iowa is only set to increase. Climate trends over the past seventy years show a steady increase in frequency of severe storms, which is likely to intensify in the next decades. Hard rainstorms will drive soil off Iowa slopes ever more rapidly, making it probable that the state average will rise and hastening the arrival of the day when swathes of Iowa's Mollisols are reduced to rubble that cannot sustain crop production.[24]

Soil erosion affects more than crop yields in the Midwest. Off-site impacts such as fouled waterways and contaminated drinking water are costly and environmentally damaging. Eroded soil and soluble nutrients are funneled from midwestern farmland into a network of rills, gullies, and streams that drain into the Mississippi River. Along its meandering trek from its origin in Minnesota through the northern agricultural region of Iowa, Wisconsin, Illinois, and Missouri, and southward through Arkansas, Tennessee, Mississippi, and Louisiana, the behemoth gathers nutrients in the water from farm runoff, or bound to soil particles, during its inexorable move toward the sea. By the time it spills into the Gulf of Mexico, the river is brown with silt and loaded with nitrogen and phosphorus. The nutrients cause explosions of algal populations, disrupting the ecosystem. What follows is far worse. The algae are photosynthetic and therefore produce oxygen, but they die and are devoured by microorganisms that guzzle oxygen at fantastic rates, thereby starving other aerobic organisms. Today, the region of the Gulf fed by the Mississippi River is among the largest oxygen-free, or hypoxic, zones in any ocean, covering an area roughly comparable to the size of Israel, Belize, or Djibouti.

The hypoxic zone in the Gulf of Mexico has devastated the local fishing industry, which is worth almost $1 billion. The accrued cost of remediation attempts by the U.S. Environmental Protection Agency (EPA) are substantial—$6.5 million in 2017 alone. Since 2008, the EPA has focused on preventing nutrients from ever entering the Mississippi River and mitigating the effects of those that do. Despite the EPA's ongoing work with farmers, tribal leaders, and universities from Minnesota to the Gulf, programs intended to reduce runoff and erosion have not shrunk the hypoxic zone. The EPA's original goal was to reduce the zone from 18,000 to 5,000 square kilometers by 2035, but by 2017, the zone had expanded to 22,000 square kilometers. One group of scientists estimated that a $2.7 billion annual investment would be needed to achieve the agency's goal.[25]

The most obvious erosion in the United States happens along coastlines and lake shores. Battering storms and rising sea levels have caused cliffs to collapse along California's coast, threatening life and property. Violent waves regularly take chunks out of Nantucket Island, the Rockaways, and beaches of mid-Atlantic states, which are receding as much as 20 meters each year. Inland, Lake Michigan reached high-water levels in 2018 and 2019. Record waves generated by the same harsh storms that erode midwestern soil are destroying fragile ecosystems along Lake Michigan's shores in such places as Indiana Dunes National Park, one of the country's top ten national parks for biodiversity.[26] The visual drama of coastal erosion makes for compelling news stories, heightening local and national concern about this exigent threat. Soil erosion in farmland, however, happens quietly and steadily. It garners less public attention, but it is just as deadly.

Crossing the equator, we arrive in South America, the continent predicted to incur the largest increase in erosion rates in the coming decades. Already 68 percent of South American soil is affected, with 259 million hectares deforested, 70 million hectares overgrazed by livestock, and as much as half the land in Argentina and Paraguay damaged by desertification. Landlocked Bolivia, where 77 percent of citizens live in degraded areas, is a particularly troubling case. Traditional farming techniques have long protected soil on Bolivia's slopes—home to the highest proportion of indigenous people in any South American country—but the trend in rural communities toward non-agricultural employment has led to labor shortages, necessitating less labor-intensive crop management practices that accelerate erosion. Although 60 percent of Bolivian drylands are eroding at 5 tonnes per hectare or less, as much as 6.4 percent of the land is losing between 50 and 500 tonnes per hectare annually. Few societies can afford losses of that magnitude, which, if left undeterred, will render those lands agriculturally obsolete in a few years.[27]

In Brazil, demand for bioenergy crops indirectly imperils the soil. Brazil has 32 million hectares of land designated as erosion hotspots because they lose soil at more than 20 tonnes per hectare each year. In the southern region, which is mixed forest and a center for poultry and pork production, erosion will likely increase as temperatures rise, leading to high-erosivity rainfall. The Brazilian economy depends on agribusiness to contribute 22 percent of the GDP and one-third of all employment. Brazil's diverse crop portfolio, dominated by beef, soybeans, coffee, and orange juice, is exported to China, the United States, and the European Union. Although agricultural expansion has boosted the nation's economy, it has been destructive to soil. Satellite imagery of Brazil shows that between 2000 and 2014 vast tracts of pastures were replaced with soybeans, sugarcane, and maize, nearly doubling the land area dedicated to intensive row crops. These three crops are responsible for an estimated 28 percent of Brazil's soil erosion caused by agricultural activity. Sugarcane production for the bioenergy industry is especially problematic because the soil is left bare and then plowed using heavy machinery that compacts land, increasing runoff and erosion. If Brazil is to keep up with increasing global demand for bioenergy alternatives to fossil fuels, farmers must restore the earth. Already, nutrient replenishment costs sugarcane farmers as much as $6 per hectare each year, and some Brazilian states are paying in excess of $200 million annually to address soil loss.[28] Heavy use of fertilizer will prop up yields temporarily, but before long, damage to soil will reduce crop production.

As soil erosion intensifies worldwide, many countries may experience crop loss simultaneously, creating unprecedented food shortages. Historically, during times of scarcity after crop failures, natural disasters, or armed conflict, countries have relied on the safety net of international food aid. But food aid programs are predicated on the assumption that certain countries will always have substantial food stockpiles. This assumption may no longer be valid.

Soil loss brings more people closer to the brink of food insecurity. With farmers abandoning an estimated ten million hectares of eroded cropland each year worldwide, the global food system is on alert. The 2014 statement by a senior Food and Agriculture Organization official who predicted that the world will run out of soil in sixty years was based on historical trends and projections. These included the loss of one-third of Earth's arable land over the past forty years, continued use of plowing, steadily rising rates of erosion in most countries, and increased frequency of severe weather. Just how likely is it that we will be able to double agricultural production in the next thirty years to feed a population of nine billion?[29]

During the twentieth century crop yields rose steadily. The Haber-Bosch process made nitrogen fertilizers readily available to farmers who could afford them. In the 1920s, scientists found that genetic crosses between unlike plants of the same species can lead to higher yields through a phenomenon known as hybrid vigor in which offspring plants produce higher yields than either of their parents. Decades of intensive crop breeding have produced plant varieties with several desirable traits such as high yield, disease resistance, and uniform plant structure that facilitated use of mechanized farm equipment. The Green Revolution of the 1960s to the 1980s extended the power of breeding, fertilizers, and irrigation to the developing world, increasing crop yields as much as 300 percent. After steady yield gains in most of the world's staple crops throughout the twentieth century, agronomists and farmers came to expect the frequent introduction of new seed varieties and management methods that would increase yield. If the past thirty years provide a good basis for predicting increases in crop yields in the next thirty, will we achieve food security in 2050? Over the past three decades, breeders and agronomists have achieved 50 percent yield increases of rice in India and maize in the United States.[30] Similar increases in all the staple food crops would bring production levels nearer to meeting the global calorie requirements, though they would still fall short of the projected need to double 2020 production by 2050.

There is, however, cautionary evidence that yields of some of the major food staples might not follow a rising trend. Worldwide, yields have plateaued in one-third of areas planted with rice and wheat and one-fourth of those planted with maize. Despite gains in rice yields under the best conditions, they have not continued to increase uniformly across the very countries whose diets are most dependent on it. On 79 percent, 37 percent, and 81 percent of the land in China, India, and Indonesia, respectively, rice yields have plateaued, and in parts of China and India, they have even declined. Likewise, wheat yields have stalled in one- to two-thirds of the agricultural land of China, India, and the United States—the three largest wheat producers—as well as in Australia and throughout the European Union. In one-fourth of farmland in France, wheat, barley, oats, and sunflower yields have all stagnated since 1990. Concurrently, wheat yields in the United Kingdom started to languish in the 1990s and have not budged since.[31] Yield plateaus are attributed to a combination of heat and drought stress caused by climate change and soil degradation leading to reduced fertility, salinization, and drought susceptibility. Although erosion is only one part of the equation, future demands on the global food system require that we optimize every aspect of crop production. If classical plant breeding or newer genetic engineering technology cannot propel crop yields through these plateaus, pressure to preserve soil rather than squander it will be intensified to meet the goal of feeding 9 billion people.

Soil erosion could impair international food aid. Throughout history people have alleviated hunger by sharing food—a tradition that forged singular alliances before food aid became a coordinated effort in the twentieth century. During the 1840s, the entire Irish potato crop suffered repeated catastrophic failures resulting from infection by the protist *Phytophthora infestans* and the onset of several years of unusually wet, cool weather. Wheat and other grains were not affected by the disease, so the British claimed remaining crops as payment of land taxes, subjecting the Irish to starvation that took one million lives. Australia, China, India, and the American Quakers all contributed food and funds. Unexpectedly, the Irish also received a donation of $170 from the indigenous people of the Choctaw Nation in the United States. The gift is especially moving because the Choctaw tribe had just been expelled from their native land and continued to struggle with extreme poverty. Perhaps their treatment at the hands of the U.S. government created a sense of kinship with the Irish persecution by the British. The Choctaw Nation's kindness was memorialized by the Irish in a monument in County Cork constructed of nine metal feathers, each over 6 meters tall, arranged in a shape reminiscent of a bowl of food. Many years later, the Irish reciprocated by sending aid to Native Americans during the 2020 coronavirus outbreak that devastated several indigenous communities.[32]

The modern history of food aid began with an international UN Food and Agricultural Organization conference in 1953, when several countries were experiencing food shortages and the United States was accumulating surpluses. In 1962 a UN resolution established the World Food Programme, which created a multilateral partnership with grain donations from the United States, Japan, and several countries in western Europe that were eventually joined by Canada and Australia. Almost every year since the 1950s, billions of tonnes of food have been donated to countries confronting shortages. These donations were critical to India and Biafra in the 1960s; the Sahel and Cambodia in the 1970s; Ethiopia and Mozambique in the 1980s; Rwanda, Honduras, and Somalia in the 1990s; Eritrea, Ethiopia, Bangladesh, Afghanistan, Georgia, and North Korea in the 2000s; and Syria and South Sudan in 2014, among many others.[33] Although food aid programs are imperfect—plagued by challenges of distribution, inefficiency, self-interest, nutritional balance, and political interference—they have saved many lives. Millions have been fed when no other food was available during droughts and floods that destroyed crops, after natural disasters such as hurricanes and tsunamis that eliminated both crops and food stockpiles, and amid armed conflict that prevented access to food.

The United Nations, which coordinates the World Food Programme, has been forced to make tough choices resulting from dwindling funds and rising needs. In 2017 the program received $6.8 billion but needed $9.1 billion as simultaneous shortages struck Somalia, Syria, South Sudan, northeast Nigeria, Ukraine, and Yemen where eighty-five thousand children died of malnutrition during three years of war.[34] With a rapidly growing world population and an increasingly stressed global food system, food aid might be wholly inadequate to avert famines during years in which drought, civil wars, and floods converge. And soil erosion is one factor that will likely limit food availability to food aid programs.

Other factors are challenging agriculture in the very countries that have been the bulwark of international food aid. Over the past seventy years, the United States has contributed 50 percent of total food aid, so its agricultural productivity has direct consequences for recipient countries. American agriculture has already suffered periods of low food stockpiles, such as in 2008, when wheat stocks were down to a twenty-four-day supply after decades of maintaining enough to last three months. The increasing frequency of severe rainstorms—those that can erode 50 tonnes of soil per hectare in a day—are undermining a long-held sense of security in midwestern agriculture. Several such storms could remove 10 percent of the soil in an average Iowa field in just one year, and if such years become the norm, the topsoil could soon be slashed in half, at which point yields would be acutely compromised. Severe storms will also cause more flooding, further reducing crop productivity. Climate change delivers extreme heat to parts of the United States, causing stress that reduces fruit development, lowers yields of several fruit and vegetable crops, and accelerates desertification across the water-deficient states of the West.[35] If U.S. food stockpiles are drawn down and more of its land is degraded, will the United States continue to provide half of all international food aid?

The situation is not necessarily stable elsewhere. Canada, another major contributor to international food aid, experiences multibillion-dollar losses each year from erosion. Several European countries that contribute to international aid have experienced alarming soil erosion trends, collectively costing an estimated $1–$20 billion of their GDP. Based on the recent trends and models that indicate continued climate warming, the Americas and Europe will be pelted with rainstorms that flood fields while drought scorches crops in Asia and Africa. These inevitable stressors of the food system mean that even in the best of years, food supplies will be thinly stretched, and prices will escalate.[36]

The convergence of soil erosion with yield plateaus, conversion of agricultural land, and a rising population creates a dim future for food sufficiency. When climate change is added to the storm, the outlook worsens. In a world connected through global food markets, climate, and conflict, every citizen on Earth should be concerned.

7

The Climate-Soil Duet

Climate and soil are intimate partners in a dance of millennia. Both may lead, both may follow. As the partners whirl around, climate remodels soil, and soil remodels climate. At its worst, the duo is destructive—soil loss accelerates climate change, and climate change intensifies soil erosion. At its best, the duo is harmonious, improving soil health and stabilizing climate. Today humans are uniquely positioned to restore the duo to harmony. The dynamics at the heart of the climate-soil duo reveals something somber but full of promise. Soil contributes to climate change, but it can also be used to impede it.

For most of the history of the climate crisis, solutions have focused on reducing carbon emissions, clean energy alternatives, protecting rain forests, and planting more trees. All are sound recommendations, but they ignore a significant source of atmospheric carbon and the largest terrestrial sink: soil. Loss of plant carbon to the atmosphere has often been discussed, but the soil carbon in the same ecosystem has rarely been mentioned. The United Nations' Intergovernmental Panel on Climate Change, which issues the definitive status of the climate every four years, devoted barely a paragraph to soil in early reports, but in 2019 it released a report, *Climate Change and Land,* which focused on land-soil interactions, desertification, and land degradation.[1] This is the moment for exploring soil's role in climate change. The issues to explore are the nature of climate change itself, the contributions of soil to greenhouse gas emissions, and the potential for soil to sequester carbon (fig. 15).

The greenhouse effect made Earth habitable for life. The planet would be frozen without the gases that surround it, trapping warmth that would otherwise be lost to space. But the amplification of the greenhouse effect by the accelerated accumulation of greenhouse gases is problematic because more gases means more trapped heat, rapidly increasing global temperatures and threatening terrestrial and aquatic life. Greenhouse gases continue to accumulate in the atmosphere and further warm the Earth's surface. The global temperature has already risen 1°C since preindustrial times, and scientists predict that an increase of more than 1.5°C will have catastrophic consequences. At the current rate of greenhouse gas accumulation, the 1.5°C threshold will be crossed between 2030 and 2052. Some regions of the planet have already been devastated by warming. Average temperatures in the Arctic rise at twice the rate of the rest of the globe, a phenomenon known as Arctic amplification, which causes polar ice caps to melt,

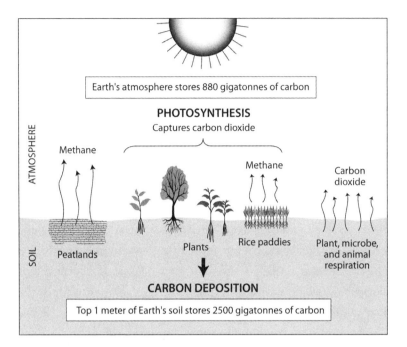

Figure 15. The role of soil processes in the global carbon cycle. Illustration by Bill Nelson.

sea levels to rise, and habitats to disappear. Rising sea levels shrink the exposed landmass, eroding coastlines and submerging islands. As marine systems heat up, biodiversity is lost in the species-rich hotspots of coral reefs and the Arctic, and fisheries are imperiled worldwide. In the tropics, climate change increases daytime temperatures, causing drought and resulting in heat stress for plants and animals. Temperate environments at middle latitudes are projected to heat up by about 3°C, causing more wildfires, desertification, and water shortages.

In 2019, all major greenhouse gases reached record high concentrations, including carbon dioxide, which reached 409.8 parts per million, the highest level on record in ice cores that encompass eight hundred thousand years. The consequences are evident—July 2019 was the hottest month documented in records going back to the 1880s, and 2020 tied 2016 as the warmest year on record. Ice cover on Northern Hemisphere lakes was seven days shorter in 2019 than the average between 1981 and 2010, alpine glaciers continued to shrink for the thirty-second consecutive year, and sea levels reached a new high.[2]

The new climate will intensify natural disasters that would have happened anyway and cause others that would not have occurred. Supercharged storms, heat waves, earthquakes, droughts, and tsunamis are becoming commonplace features of Earth's weather. A snapshot of 2020's weather reveals floods in Pakistan and Greece; several hurricanes over the southeastern United States, including one with unprecedented windspeeds; severe drought in the western United States; and the highest rainfall in twenty-four years in Jakarta, following one of the worst droughts recorded in Indonesia. Over the next century, the frequency of extreme conditions will likely rise to unprecedented levels. The International Panel on Climate Change projected that by 2100, the Earth will experience a 0.3°C–4.8°C increase in surface temperature. The warming will be accompanied by a rise in sea level of between 20 and 100 centimeters. These changes are predicted to be accompanied by increases of 10–30 percent in average precipitation across many parts of the world and decreases of a similar magnitude in others.[3]

Climate change increases flooding several ways. As land warms, more water evaporates from the soil and becomes fodder for storms. Warmer air holds more water, giving it more to dump in a rainstorm, and more extreme rainstorms deliver water more quickly than it can be soaked up by the earth, causing flooding. A hotter atmosphere and warmer oceans can increase the speed of tropical storms, sometimes escalating them to hurricanes. As sea levels rise, more areas become vulnerable to severe flooding. Researchers projected that a 0.5-meter rise in sea level, which is likely to occur by 2060, would double the tsunami frequency in Macau. A 1-meter rise, predicted to happen by

2100, would increase the hazard by up to 4.7 times. In the large deltas where major cities in Asia reside, the risk of flooding is set to rise drastically over the next fifty years. Considering projected population growth in many of these urban centers, an estimated ten times more people will be vulnerable to such floods by 2070.[4]

Natural disasters place a strain on the global agricultural system, reducing food production and increasing the need for humanitarian aid. Between 2001 and 2016, the United States slashed its budget for nonemergency food aid in half and increased emergency aid by 50 percent. This period included some of the worst natural disasters in modern history. The 2004 tsunami in Southeast Asia took over 200,000 lives and displaced 1.7 million people from their homes. The World Food Programme, to which the United States is the largest contributor, provided 169,000 tonnes of food to 2 million people at a cost of $185 million. After Hurricane Mitch, the United States provided $67 million worth of food aid to Honduras. Although the U.S. budget for food aid between 2006 and 2016 was $2.5 billion per year, there is an estimated global shortfall of $1.3 billion for emergency food. If the frequency of natural disasters increases twofold to fourfold, as predicted under climate change, will the United States have sufficient excess food to cover global needs during emergencies? Recent patterns do not bode well. In 2019, fourteen catastrophic weather events, including ten severe storms and tropical cyclones, three floods, and one wildfire, collectively cost $43.5 billion.[5]

People across the globe will feel the repercussions of climate change, but inequalities within and among countries will unevenly distribute the burdens. People in poorer countries are four times more likely to be displaced from their homes by extreme weather events than those in wealthy countries, and 80 percent of those displaced are women.[6] Climate changes will exacerbate ongoing challenges faced by indigenous people. Droughts in the Kalahari Basin in Africa are already forcing indigenous people to live near government wells to obtain water. Bangladeshi farmers have started floating gardens for growing vegetables during floods. Many people in the Arctic face a precarious future as populations of the animals they depend on for food diminish. In the Himalayas, indigenous people will lose their water supplies as snow cover shrinks and glaciers at high altitudes melt and then dry up.[7] As climate change progresses, many societies will be remodeled to accommodate Earth's changing surface.

Climate change also affects people through its effects on other organisms. It inflicts new challenges on crop production by altering the behavior and habitats of numerous plant pests and pathogens. Coffee leaf rust epidemics illustrate the interconnectedness of climate and plant disease. Rust is the fungal disease that destroyed coffee production in Sri Lanka in the 1880s and impelled the farmers to instead plant tea. Ostensibly, this disease is responsible for the British taste for tea because it destroyed coffee plantations in British colonies. Over most of the twentieth century, the leaf rust spread throughout the world, but it was excluded from South and Central America by stringent quarantine measures. In the 1970s the fungus broke through the barricade in Brazil and spread throughout Latin America. For years, the safe haven for coffee trees was land above 1,000 meters elevation, where the fungus died from cold nighttime temperatures. As climate change has increased air temperatures, regions that were free of coffee rust are now infested, and coffee plantations have been forced to move to higher altitudes.[8]

Climate warming has expanded the range of such insects as the voracious southern pine beetle, which has made a pilgrimage north, denuding trees and reducing thriving forests in the United States and Canada to expansive graveyards with towering skeletons of pine trees commemorating the dead. In 2020, the locust infestation in East Af-

Plate 1. Left, Entisol soil profile; right, Spodosol soil profile. Photo courtesy of the USDA—Natural Resources Conservation Service (left); Dahlhaus Kniese / Alamy Stock Photo (right).

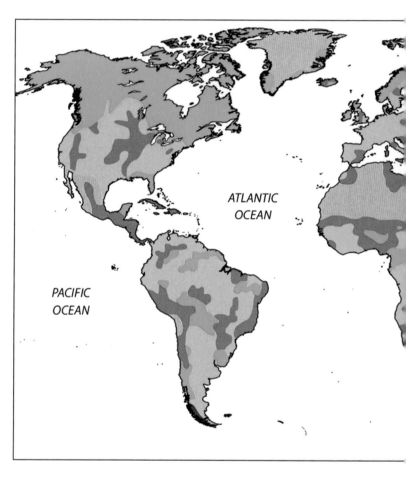

Plate 2. World map of soil degradation. Illustration by Bill Nelson and based on global map of degraded soils using data from the GLASOD study in 1997. Philippe Rekacewicz, UNEP/GRID-Arendal, https://www.grida.no /resources/7424.

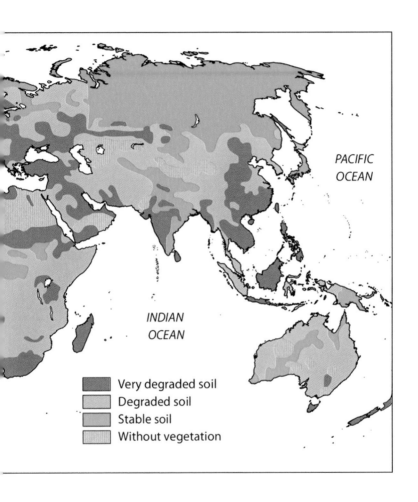

PACIFIC
OCEAN

INDIAN
OCEAN

Very degraded soil
Degraded soil
Stable soil
Without vegetation

Plate 3. Eroded sediment in the Waiapu River in New Zealand, 1999. Photograph by Dr. Noel Trustrum, courtesy of Landcare Research NZ Ltd., 1988.

Plate 4. Left, the Rhône River carrying sediment, flowing from Valais into Lake Geneva, Switzerland; below, urban gully in Kinshasa, Democratic Republic of the Congo, April 2019. Photograph by Rama, Wikimedia Commons, Cc-by-sa-2.0-fr (left); photograph by Matthias Vanmaercke (below).

Plate 5. Top, a recently planted field in the Chernozem region of Ukraine: Chernozems are some of the deepest soils in the world, carrying up to 1.5 meters of topsoil; bottom, severe erosion in Ukraine. Photograph by Anton Petrus (top); photograph by Yurikr Megapixl.com (bottom).

Plate 6. Top, erosion in northern Iowa reveals sandy knobs (light-colored parent material) where the topsoil is entirely eroded; bottom, strip cropping with prairie plants in Iowa. Photograph by Lynn Betts, courtesy of the USDA (top); photograph courtesy of Omar de Kok-Mercado, Iowa State University (bottom).

Plate 7. Terraced rice fields in Mu Cang Chai District, YenBai Province, Northwest Vietnam. Photograph supplied by Tzido—Freepik.com.

rica provided yet another reminder of climate's power to regulate—or deregulate—pest populations. A local infestation spread to countries across Africa, the Middle East, South America, and South Asia.[9]

As climate change drives soil off the land and intensifies disease, pest damage, and extreme weather events, the gathering storm is unmistakable. It is hard to envision global crop yields keeping pace with the needs of the burgeoning population. All citizens of planet Earth will feel the effects of anthropogenic dominance on the natural world, but inhabitants of coastal areas, indigenous people, and those living on low-altitude islands will be especially hard hit.

Soil contains 2.5 trillion tonnes of carbon, which is triple the weight of Mount Everest, making it the largest carbon repository on terrestrial Earth. Although public discourse about climate change focuses on atmospheric carbon, soil contains three times more carbon than Earth's atmosphere and quadruple the amount stored in vegetation. It is this precious stock of immobile carbon that places soil and climate in a delicate balance that, if disrupted, can spin out of control with extreme repercussions. Sometimes erosion can bury soil, storing its carbon safely underground. Other times erosion mobilizes soil carbon, some of which is converted to greenhouse gases. Simultaneously, eroded soil's capacity to support photosynthesis is diminished, compromising the most important counterbalance to atmospheric accumulation of carbon dioxide. Thus, the vicious cycle continues: climate change accelerates erosion, and erosion accelerates climate change.[10]

Erosion is only one of the ways that agriculture generates greenhouse gases. Food production is responsible for 24 percent of anthropogenic greenhouse gas emissions; the other 76 percent is the result of industry, transportation, and domiciles. Ever since the beginning of agriculture, people have been replacing natural ecosystems with cultivated farmland, blasting 133 billion tonnes of carbon from land into the atmosphere—*133 billion tonnes.*[11] Collectively, activities such as clearing and burning forests, draining soil, and expanding farmland create a steady stream of carbon dioxide, methane, and nitrous oxide into the atmosphere. Methane and nitrous oxide are especially vexing because they are high-potency greenhouse gases that are boosted by wetland rice, cows, and application of nitrogen fertilizers—sources of pollution slated to increase over the next few decades as we try to satisfy human food demands and preferences.

In addition to erosion, components of soil are lost through the process of volatilization of organic matter to methane and carbon dioxide. It is a normal, indeed necessary, process by which matter is cycled from soil into the air, but accelerating it depletes soil carbon and bloats the atmosphere with greenhouse gases. In contrast to erosion, which announces itself through its signature gullies and swirling dust storms, as well as other forms of pollution that assault the nostrils, blacken the sky, or discolor rivers, volatilization is a covert operation.

In June 2020, a sandstorm transported an unusually large amount of the Sahara Desert to the atmosphere and eventually to the Western Hemisphere. Although the Sahara loses 800 million tonnes of sand every year, sprinkling much of it over the Americas, the 2020 mineral migration was historic in density and size. For days, the unhealthy air and flaming sunsets were all over the Western news. In contrast, every year 60 billion tonnes of material from soil are turned into carbon dioxide and other gases that float unnoticed into the atmosphere.[12]

Microorganisms are the culprits behind soil volatilization. They convert soil matter into the invisible greenhouse gases carbon dioxide (CO_2) and methane (CH_4) by decomposing organic matter, and into nitrous oxide (N_2O), the product of combining excess fertilizer with oxygen. Between 2007 and 2016, agriculture and forestry produced 13 percent of the carbon dioxide, 44 percent of the methane, and 81 percent of the nitrous oxide released from all sources on Earth, anthropogenic and otherwise.[13]

The most abundant greenhouse gas is carbon dioxide. In soil it is produced by animal and microbial respiration, a process used to extract energy from food. Although respiration and breathing are used interchangeably in colloquial lexicon, respiration actually refers to a set of biochemical reactions for processing food into energy. Respiring organisms include those that can breathe in the sense of actively inhaling and exhaling air, as well as those that do not breathe but nonetheless use oxygen as fuel. Soil animals and microbes gobble up many carbon compounds they encounter, pumping out carbon dioxide as they go. A surprising secret about respiration is that plants do it too! Although their most celebrated metabolic activity is the *fixation* of carbon and the *production* of oxygen through photosynthesis aboveground, below ground it is a different story. The carbohydrates that plants transport to roots are metabolized with oxygen to produce energy, reversing the photosynthesis equation, nourishing roots and soil, and releasing carbon dioxide as a waste product.

In the absence of oxygen, certain soil microbes can switch to a different way of making energy—anaerobic carbon metabolism. Anaerobic organisms consume many different carbon-containing substances from which they produce methane or, to a lesser extent, carbon dioxide, as waste products. This is the same type of bacterial metabolism that produces massive amounts of methane in the rumen, the upper part of the digestive tract of ruminant animals. Anaerobic metabolism is sluggish compared with the brisk pace of its aerobic counterpart, and so anaerobic environments accumulate carbon matter. Wetlands are well known for their anaerobic conditions and carbon accumulation. Composing only 7 percent of the world's land area, they store 25 percent of soil carbon worldwide—equal to the amount of carbon in the atmosphere. Most wetland carbon remains in the soil for millennia, but a portion is processed by anaerobic microbes known as methanogens, which produce methane. Although wetland methanogens convert only a small proportion of the carbon, they contribute one-fourth of the total methane produced on Earth— between 100 and 250 million tonnes annually.[14] Much of this conversion occurs in Histosol peat bogs and fens.

Histosols are a specialized type of wetland soil that constitute approximately 1 percent of global ice-free land. They are famous for their wet, anaerobic conditions that produce an extremely deep organic layer. Centuries of high rainfall saturates the soil, replacing air with water, creating the anaerobic conditions that turn these patches into the peat bogs and fens of Europe, North America, and Southeast Asia. Plants with roots adapted to boggy conditions deposit carbon in the soil faster than it is consumed by the sleepy anaerobes, amassing a deep stock of carbon and a dense fabric of organic matter, which gives peat its characteristic dark color.

The slow, steady release of methane is a normal part of life in a bog, but if oxygen is added to the system, then volatilization of carbon and nitrogen compounds vastly accelerates, turning the bog from a carbon sink into a source of carbon dioxide. This process is occurring in the deep peat bogs of the United Kingdom, which can contain as much as 10 meters of peat—the result of thousands of years accruing carbon. Today, erosion, fire, pollution, and drainage disrupt the balance of plant photosynthesis and microbial activity, reducing carbon deposition and revving up aerobic metabolism that converts stored organic matter to carbon dioxide. Peat is also harvested in rural areas for home-heating fuel and commercially for the horticultural industry to enrich garden soil. In regions where peat is mined for fuel or compost or where dry conditions are needed for construction, the bogs are drained, converting them to aerobic, erodible environments. Once the bogs have been stripped of plants and water, erosion takes command, gouging gullies in the landscape. Although they represent less than 10 percent of bog land area, gullies produce much of the carbon dioxide emitted from the United Kingdom's peat bogs. This land that was once the custodian of a massive carbon sink now releases carbon dioxide at a rate of 3.7 million tonnes per year, roughly equivalent to the annual emissions of 700,000 British households. The Earth's entire complement of peatlands, found predominantly in Asia and North America, is emitting a thousand times this amount.[15]

Asia is home to one-third of the world's peatlands, some of which are being volatilized rapidly. In Malaysia, the oil palm industry competes fiercely with efforts to protect peatlands. Oil palm is the most productive oil crop in the world, producing 3.3 tonnes of oil per hectare, which is almost five times more than coconut, sunflower, or canola and eight times more than soybean. The palm industry has expanded into the peatlands of Malaysia's state of Sarawak with the stated intention of meeting global needs for oil used in food processing, cosmetics, animal feed, and biofuel while providing local employment and reducing rural poverty. To maximize oil production on the plantations, the peatlands have been drained. As the water is bled into pipes, air rushes in to fill newly empty pockets in the soil, and aerobic microbial activity surges. The rapid pace of aerobic metabolism results in accelerated decomposition of the organic matter. The peat disappears at an astonishing rate, losing as much as a meter (20 percent of its initial depth) in just seven years, releasing 15 tonnes of carbon per hectare in the first year and rising to 18 tonnes per hectare annually by the seventh year after drainage. Across Southeast Asia, drainage of the peatlands collectively emits about 1 gigatonne of carbon into the atmosphere annually—twice the weight of all people on earth.[16] Intact wetlands are the largest natural contributors to global methane, but agriculture and forestry augment their carbon contribution.

Flooded rice paddies are cultivated wetlands whose methanogens spew 18–39 million tonnes of gas into the atmosphere annually, accounting for 20 percent of anthropogenic methane sources. Rice, a semiaquatic plant, feeds more than half the world's population, and demand for it will likely increase 20 percent by 2035, triggering a surge of methane production unless mitigation measures are adopted. Research shows that mitigation is possible. In oxygen-free rice-paddy soil, organic matter is converted to methane, and in highly oxygenated soil, carbon dioxide is produced. But at a middle level of oxygen, little of either gas is produced. Scientists are pursuing ways to manage oxygen availability in rice-paddy soil to achieve this ideal condition. If successful, this work could diminish the environmental impact of rice farming, which is currently an acute threat to climate.[17]

Mitigation of methane production in rice fields may also happen as a consequence of scientists' attempts to increase rice yields. In an auspicious twist of biology, microbes associated with high-yielding rice varieties release less methane. This happens because higher-yielding varieties transport more oxygen to their roots and release it into the rhizosphere, where it stimulates microbes to use methane as a food source. Microorganisms that eat methane, or methanotrophs, carry out the opposite chemical reactions as methanogens, converting the potent greenhouse gas to harmless biomass. Similarly, adding straw to paddy soil increases yield in the short term and reduces methane emissions in the long run by boosting growth of methanotrophs. Draining fields introduces oxygen, which suppresses methane producers and rouses methane consumers, but care must be taken not to cross the threshold of aerobic conditions that favor carbon dioxide producers.[18] Insight into methane production in rice fields offers a glimmer of hope that we may be able to increase production of our most prevalent food crop while reducing its impact on climate change.

In Gelisols, the permanently frozen soils, methanogens are metabolically quiescent because cold stills the chemical reactions needed to produce energy while keeping the microorganisms alive. When Gelisols warm, dormant microorganisms become active and begin exhaling methane. This process has been detected across the Northern Hemisphere in alpine regions that are thawing for the first time in millennia, and some sites are not refreezing, even in winter.[19] As Earth's icy soils melt, more methanogens will wake up and expel the gas that becomes an atmospheric heatwrap around the Earth. And thus, another vicious cycle ensues: global warming stimulates production of soil methane, intensifying the greenhouse effect, which causes more thawing and incites erosion.

Methane receives lots of attention as the infamous villain of climate change, but an even worse actor is nitrous oxide. This insidious gas has ten times the warming potency of methane and three hundred times that of carbon dioxide, and it is frighteningly long-lasting in the atmosphere. In addition to its bad behavior as a greenhouse gas, nitrous oxide destroys the ozone layer, Earth's ultraviolet shield, and has now surpassed chlorofluorocarbons as the top ozone-destroying gas emitted from our planet. Nearly half of the nitrous oxide produced as a result of anthropogenic activity is emitted by microorganisms in soil, and the rest is from microorganisms in manure.[20]

Nitrous oxide emissions doubled between 1961 and 2016, largely due to an eightfold increase in the use of nitrogen fertilizer. Crops take up only 50 percent of the applied nitrogen, and as excess accumulates underground, nitrous oxide-producing bacteria turn the soil into a pumping station for the gas.[21] Fortunately, the same moderate oxygen levels that minimize carbon dioxide and methane production in rice soil also result in little nitrous oxide. Thus, at least in rice production, there may be a good strategy for managing nitrous oxide production. To get the most out of their fertilizers and lose less to the atmosphere, many farmers carefully formulate, place, and time nitrogen fertilizer applications to maximize uptake by plants rather than by hungry microbes. But the prevalence of atmospheric nitrous oxide and nitrogen fertilizer runoff in lakes and rivers leads to the inescapable conclusion that nitrogen has been overused and mismanaged on agricultural land. We can do better.

If erosion is accelerating climate change, can soil and its photosynthetic inhabitants also slow it down? Put simply, when carbon is in the soil, it's not in the atmosphere. Slowing conversion of soil carbon to greenhouse gases would therefore be expected to reduce the greenhouse effect. Fewer energy-absorbing molecules in the atmosphere will lighten the blanket that traps heat around the Earth, enabling more radiant energy to leave the atmosphere and dissipate in outer space rather than warm the surface of the planet.

If the biochemical reactions that maintain life were linear, with a definite beginning and end, then all biological molecules would pile up at a dead end. If this were the case, then the evolution of life on Earth would not have advanced very far because the planet has a finite number of atoms. Instead, the metabolic processes that produce energy are part of cycles, and cycles are bidirectional. If the collective metabolism of soil-dwelling organisms drives carbon from soil into the atmosphere, there must be a way to accelerate the cycle in reverse to sequester atmospheric carbon in the soil. And there most assuredly is.

There have long been theories about how climate change itself might increase photosynthesis, thereby reducing atmospheric carbon dioxide. In the past two decades, these theories have been substantiated with many studies, including those examining the Sahel, the semiarid band of countries in Africa that serves as a transition zone between the Sahara Desert to the north and the humid savannas to the south.[22] The region extends from coast to coast across ten countries, from Senegal on the west coast to Sudan on the east. Its 7 million square kilometers are home to almost 135 million people who have sustained themselves for centuries with crop and livestock agriculture despite the harsh conditions. Food production and control of trans-Saharan trade routes made the Sahel a bustling center of commerce in north-central Africa more than a thousand years ago. But by 1980, stretches of the Sahel would have been unrecognizable to those who crossed it in the nineteenth century because the northern part was no longer on the edge of the desert—it *was* desert. The Sahara had marched south to engulf part of the Sahel.

This was a consequence of unusual weather patterns coinciding with overgrazing, which led to loss of vegetation and soil. Less plant cover increased albedo, or surface reflectance—a property of the Earth that cools its surface. In the Sahel, this cooling effect reduced rainfall. Compounded by the effects of longer-term climate trends, an extended drought ensued, peaking in 1973. The Sahel had suffered droughts before, but not of this length or severity. Between 1969 and 1978, water scarcity decimated cattle and killed approximately one hundred thousand people in the region's densely populated countries.[23]

In a remarkable turn of events, much of the Sahel is now green again, due to an unexpected benefactor: climate change. Since the 1980s, rainfall has increased in the region, and the southern border of the Sahara is again creeping northward. Satellite images are reinforced by people on the ground who once again are able to farm in regions they were forced to abandon during the drought. Part of the greening is due to increased rainfall, which modeling predicts will increase as much as 1–2 millimeters per day during the three-month monsoon between 1980 and 2080. This is certainly part of the yield boost. But in the thirty years following the worst of the drought, millet yields increased in Burkina Faso by 55 percent and in Mali by 35 percent, more than can be explained by increased rainfall alone.[24] Increasing yields are thought to be caused by rising carbon dioxide concentrations in the atmosphere.

Enhanced atmospheric carbon dioxide levels appear to have accelerated photosynthesis up to 33 percent over the past three decades in China, India, North America, southeast Brazil, southeast Australia, and parts of Europe, burying an extra 6 billion tonnes of carbon in soil between 2007 and 2016. As climate change lengthens growing seasons in some locations, plant productivity increases. But let's be realistic: photosynthetic acceleration can only partially offset increasing emissions.[25] Soil can provide further assistance.

Soil's global potential for additional carbon sequestration is between 1 and 3 gigatonnes per year. Over the next few decades, a 10 percent increase of organic matter in soil worldwide could reduce atmospheric carbon dioxide concentration by 25 percent, or 110 parts per million, nearly returning to preindustrial levels.[26] And there is a double benefit—increasing carbon in soil would simultaneously restore its health. Despite the clear benefits and little accompanying risk, scientific and policy discussions have focused attention elsewhere.

Increasing soil carbon requires simple inputs that enrich the soil ecosystem. Breeding plants that spill more of their carbon into the soil along with alternative planting practices will increase soil carbon, but these are useful only if the carbon remains in the soil for a long time. Only one-third of crop residues remain in soil after a year and only 5–10 percent remains after two years; therefore, to extract the full benefit of innovations that increase soil carbon, they must be coupled with strategies to stabilize it.

One promising technique for storing stable carbon is to enrich it with biochar, an organic material that has been subjected to high temperatures without oxygen. Currently biochar is being tested as a means to increase soil carbon storage and to improve plant growth. Biochar is more stable than soil organic matter, but the jury is still out on whether it slows conversion of soil carbon to atmospheric carbon dioxide. However, biochar most definitely has a future in agriculture because it is emerging as a one-stop shop for soil improvement. It diminishes nitrous oxide emissions, reduces fertilizer demands, and increases the water-holding capacity of soil.[27]

Wetlands are considered a prime possibility for increasing carbon stores. Several countries have implemented programs to restore anaerobic conditions in bogs to prevent further oxidation and build up their carbon stocks. Indonesia is home to 36 percent of the world's peatlands, containing 28 gigatonnes of carbon. Peat is found in Indonesia's publicly owned forests, which have become a target for commercial development. In the past, companies acquired rights to forest sections from the government and burned them to develop rubber plantations, releasing large amounts of carbon into the air. The problem was compounded by agribusinesses that installed pipes to drain wet areas, increasing the soil's susceptibility to fire. In dry years, massive conflagrations swept the peatlands, forcing schools to close throughout Indonesia and Malaysia and preventing planes from flying through the smoke-filled skies. In 1997 and 1998, fires in Malaysia released almost 1 gigatonne of carbon into the atmosphere, increasing human mortality by as much as 20 percent.[28] The peat is so deep that the fires could burn for years if they weren't quenched by the annual monsoon.

Peat fires can be prevented. After a bad fire season in 2015, Indonesia's government established the Peatland Restoration Agency, tasked with restoring 2.6 million hectares of degraded peatlands and preventing peat fires. It has since banned removing forests and setting fires. The restrictions seem to be working—in 2019, another dry year, the burned land area was reduced 87 percent compared with 2015.[29]

Reducing agriculture's contribution of greenhouse gases to the atmosphere would improve climate, but people need to do the hard work to make it happen. The majority of anthropogenic sources of greenhouse gases are the result of burning fossil fuels for transportation, heating, and industrial processes. Mitigation efforts must focus on reducing these emissions. Agriculture is also an important contributor to solutions, accounting for 24 percent of emissions, so curtailing carbon dioxide, methane, and nitrous oxide production from soil and livestock would have a modest effect. But the big payoff could come from incorporating more carbon into soil and stabilizing it to reduce greenhouse gases. Most countries have found it challenging to reduce their emissions, but increasing soil carbon might be an antidote to fossil fuel combustion while people search for and implement energy alternatives.

At the Twenty-First Conference of Parties to the United Nations Framework Convention on Climate Change in 2015 in Paris, known as COP 21, world leaders agreed to the historic Paris Agreement, which provided a blueprint for slowing global warming. Ratified by 125 countries, this agreement aimed to limit temperature increase to 1.5°C over preindustrial times. The ambitious plan was rightfully heralded as a landmark in efforts to manage climate responsibly. A far less celebrated agreement introduced at the COP 21 summit was the "4 per mille Soils for Food Security and Climate," or 4p1000, which proposed that all countries increase carbon content of the top 2 meters of soil by 0.4 percent per year. The calculation was predicated on the goal of offsetting future increases in carbon emissions, which are projected to be equivalent, annually, to 0.4 percent of the carbon content in the world's soils. Many scientists wondered whether it was possible to achieve this goal and in particular how much carbon could be incorporated into nonagricultural soils. A few months after COP

21, an international group of soil scientists proposed a more feasible goal—to increase soil carbon by 0.4 percent in the top 1 meter of soil on agricultural lands. Success would mean sequestering 0.4–3.0 gigatonnes of carbon per year.[30] Although it would not cancel out all emissions, accomplishing 4p1000 on agricultural land would reduce them by up to one-third. A U.S. president dedicated to restoring climate discussions to a prominent stature could resuscitate, ratify, and implement 4p1000 to advance the world's climate mitigation and improve soil.

The difference between actual and potential carbon content is known as the soil health gap. Filling this gap would require incorporating about 133 gigatonnes more carbon into the world's soil. Even if just 0.5 gigatonnes of carbon was sequestered each year, by 2100, the atmosphere would have been spared 40 gigatonnes of carbon, which is equivalent to 15 percent of the anthropogenic carbon dioxide emitted from 1750 to 2015.[31] Working toward the 4p1000 aspirational goal seems worth exploring throughout the world until the carbon-sequestering potential of soil is used up, which will likely take several decades. By then, alternatives to burning fossil fuels may be widely available, and new mitigation strategies may be implemented. In the meantime, increased carbon would also benefit soil health, which would eventually help farmers enhance yields and reduce inputs.

We can interrupt the destructive dance between soil and climate now. Worsening climate and continuing soil loss create a vicious cycle that whirls toward destruction. In the face of the climate change threat, the earth may be our best ally. The methods for building soil health are well known, having been practiced for centuries. Wise food producers around the world are already intent on restoring harmony by reversing the carbon cycle. Good soil stewardship is possible even in the face of feeding nine billion people. More than possible, it is necessary.

8

Soil Stewards

Beneath the bustle of cities, towns, farms, forests, and highways lies the silent, dark ribbon of life, rock, and water that binds the past and future. People living in the present hold the power to strengthen or break that bond. The ribbon chronicles our actions, baring them for the future to see. During our brief tenure as custodians of soil, we may nourish it, leaving a better legacy than the one we inherited. Or with neglect and abuse, we fray the ribbon, creating an uncertain future for generations to come and the Earth itself.

Soil legacies often find their origins in cultural values and beliefs. Some cultures treat soil as merely a means to a productive and profitable end. Others honor it as a divine maternal source of all life. For many, soil seems renewable and limitless, while others view it as a vanishing resource that must be protected. Whatever their beliefs, people throughout history and across the world's cultures have left soil legacies that tell stories. Examining the practices that created strong legacies teaches us how to nurture today's soil so that it lasts until tomorrow. A powerful theme emerges: we know how to do this! For millennia farmers have managed wind, water, and the soil itself to minimize erosion. The practices were refined, mechanized, and quantified in the last century but are not new to the modern age of agriculture.

A comparison of soil management across cultures illuminates a second theme: good soil stewards converge on the same management methods whether they live in the Northern or Southern Hemisphere, grow maize or tef, use planting sticks or spades. Their continued practice of agriculture today is proof that long-lived agrarian societies tend their soil with care. Those that squander soil simply do not survive.[1]

When I was first learning soil science, I was overwhelmed by the variety of soils and soil management practices. Wanting to see them all, I imagined having the superpower to fly over the world's landscapes and peer into the underground activity influencing soil health. Today my youthful fantasy is well approximated by a drone or satellite equipped with laser sensors, although drones don't usually accept passengers! But if we could hitch a ride on one of those drones, we would see some astonishing examples of land stewardship—places where humans have cultivated soil richness that far exceeds what was there originally. The resulting soils are Anthrosols, shaped by anthropogenic activity. For example, flying over the tiny island of Papa Stour off the coast of Scotland, we would still see the results of the plaggen agriculture long practiced on many Shetland Islands, first by Norse settlers around 800 CE and then by the Scottish people. On Papa Stour, agriculture continued until the 1960s, when the island's population declined, reducing the need for food production. Much of the plaggen-influenced agricultural land on Papa Stour has remained uncultivated for the past sixty years, enabling curious scientists and anthropologists to reconstruct the practices that created its signature Anthrosol. For over one thousand years, farmers nourished the land with copious manure, turf, seaweed, and byre muck (the bedding from cow stalls) and worked the soil with spades rather than plows. When they finished farming, the land chronicled a 75-centimeter legacy of topsoil compared with only 16 centimeters in uncultivated surrounding areas. Most farmers in western Europe abandoned plaggen agriculture when the Haber-Bosch process made nitrogen fertilizers readily available, and today their topsoil has been substantially reduced. But the Papa Stour Anthrosol remains as a testament to the ancient practice.[2]

Underground monuments memorialize the lasting benefits of other past practices. Under the dense forests of South America, for instance, scientists found the soils known as Amazonian Dark Earths. In fact, scientists discovered this Anthrosol in pockets of the Amazon rain forest using a laser much like the one on our fantastical drone.[3] These black soils are found in places that were once home to small communities of indigenous people who raised plants for food in the forest. The Amazonian Dark Earths contain deep stores of organic matter deposited by cool burns that produced biochar. People enriched the soil further with plant waste dropped in the town dump, which generated great compost. The residue of biochar and compost still freckles the forested landscape. Its richness contrasts sharply with the surrounding thin tropical soil, making it easily detectable by lasers even centuries after the land was abandoned to the jungle.

If our drone took us northward from the Amazon to Xochimilco, just south of Mexico City, we would catch sight of chinampas, or artificial floating gardens, which some historians believe enabled the Aztecs to build their great empire. The farming technique traces back to pre-Aztec Mesoamerica between 800 and 1000 CE; later the Aztecs launched massive canal construction projects, expanding chinampa food production to meet the needs of their population of five to six million. Chinampa technology hasn't changed much in over one thousand years—mud from the bottom of lakes and canals is combined with peat from land to build up piles of the jet-black mixture that reach from the bottom of the waterway to the surface. Although they are still referred to as floating, the islands are in fact anchored to the bottom of the waterways. The surfaces of the islands display a dark Anthrosol that is held firmly in place by vegetation bordering the islands, preventing erosion into the canals. Only 10 meters wide but stretching several kilometers, these long, skinny islands carve the canals into narrow channels. The artificial islands are a remarkable example of human ingenuity. The soil on the chinampas continues to support productive agriculture—farmers raise fruits, vegetables, and flowers throughout the 406 kilometers of canals. The islands add 2,000 hectares to Mexico's farmland, supplying organic produce for twelve thousand people in Mexico City and supporting 11 percent of Mexico's biological diversity.[4]

Moving eastward to the Philippines' Cordillera Central region on the island of Luzon, we would see just over 17,000 hectares of land, much of which has been farmed for two thousand years by the Ifugao people. They continue to manage their land with a traditional, highly structured system of forests, farms, rice terraces, and settlements. The land flaunts lush terraces that encircle mountains too steep to farm without recontouring (see pl. 7 for terraces). The Ifugaos' farming methods mold this rugged terrain, enabling it to withstand the high-intensity 320 centimeters of annual rainfall it receives. Only adherence to long-tested methods to reduce water velocity and trap sediment enables them to keep erosion at 0.068 tonnes per hectare per year—a remarkable feat, given that nearby unterraced land erodes on average more than 24 tonnes per hectare per year.[5] Thousands of years of successful agriculture attest to the effectiveness of the Ifugao system for building a stable agroecosystem on carefully tended soil.

Farming outcomes are typically invisible because they lie beneath the opaque skin of the Earth, but sometimes remote sensing detects differences in surface color or elevation that indicate deeper impacts. These ancient soil relics reveal a treasure trove of innovations that improve soil quality and prevent its loss. Let's now end our drone flight of fantasy and turn to the task of identifying scientific elements inherent in great soil stewardship.

There are two universal principles in preventing erosion: first, manage the forces that move soil; second, improve soil structure. Since wind and water drive erosion, any intervention that reduces their velocity also reduces erosion, as do the Ifugaos' terraces. Just as important as managing the forces that transport soil is building the soil from within—improving its architecture by binding separate particles into aggregates that are more difficult to propel. Aggregation is promoted by deep-rooted perennial plants, cover crops, and crop rotations, all of which enhance the soil's microbiome and organic matter.[6] As the Amazon forest and Papa Stour illustrate, farmers can create productive Anthrosols with crop residue, manure, biochar, compost, and other amendments that increase soil organic matter, which in turn promotes aggregate formation, reducing erodibility. And good soil stewards protect soil structure by minimizing plowing.

Where did these practices originate, and what can we learn from them? Probing the methods that have enabled people to triumph over erosion since the beginning of agrarian life builds an appreciation of the diversity, simplicity, and power of soil-protective farming practices. Indigenous knowledge has been subjected to the harsh selective forces of the world, a process of sifting and winnowing to continue practices that work and discard those that don't. The body of knowledge held by the world's 350 million indigenous people is much larger than their population.[7] The details of indigenous practices offer complex, nuanced lessons for all agriculturalists.

Given their many achievements, it should be no surprise that the ancient Maya developed sophisticated methods of soil stewardship. For nearly four thousand years Maya civilization sprawled across the Yucatán Peninsula on land that includes parts of modern-day Mexico, Guatemala, Belize, El Salvador, and Honduras. The population is thought to have declined after 900 CE, and the Maya battled colonial invaders for years after the Spaniards' arrival in 1502, but tenacious Maya farmers continue their traditional agricultural practices today. They are both inspiring and enlightening.

Scientists have gleaned knowledge of the ancient Maya from archaeology and from the soil itself. Murals decorating ancient walls and ceilings still flaunt vivid red, turquoise, orange, brown, and green pigments depicting in exquisite detail Maya government, religion, and celebrations that inform us about their agriculture. The Maya also left behind records in intricate hieroglyphics that still have not been fully deciphered. They applied their renowned arithmetic prowess and study of the cycles of the natural world to create a calendar system to synchronize agricultural operations around the wet and dry seasons. They followed the solar cycle with technology that captured the angle of the sun, a challenging task in a nearly equatorial location.[8]

The abundance of Maya agriculture is indicated by evidence of stockpiles and a social system to distribute excess. Their cropping system centers around maize, which was domesticated from a wild plant in the lowlands of western Mexico around 7000 BCE.[9] In the case of maize, we have clear genetic evidence of the domestication process, which involved selecting plants that conformed to the agricultural system and yielded useful food. High-producing maize supported development of settlements, villages, and cities along with food stockpiles. The Maya included as many as seven other plant species in their cropping system, which was part of the secret to their continuous productivity and soil protection. Maya agriculture enabled rapid growth of the population and emergence of their complex, sophisticated society.

Much has been made of the Maya collapse, a purported decline in population from six million to five hundred thousand starting in 900 CE. The reasons for the decline remain unclear, and like so many gargantuan historical shifts, it was likely caused by several factors that undoubtedly differed over space and time. Analysis of tree rings and radioisotopes suggests that the Maya were confronted with prolonged drought around the year 900. Some historians have documented a concurrent reduction of food stockpiles, suggesting that agricultural production declined with water availability; others dispute this and say that there is no evidence of food shortages. A 2020 study reported high levels of mercury in the reservoirs surrounding the palace and temples in Tikal, once a Maya metropolis. Along with the mercury, the reservoirs also contained DNA signatures of cyanobacterial species—producers of potent toxins that still plague water systems today. Contaminated water might have contributed to diminishing the population. Some scholars say that there was never a precipitous decline but rather a gradual change over hundreds of years accompanied by changes in climate, governance, and societal organization. Yet despite all of these possibly contributing factors, soil erosion has been blamed for the decline of food abundance and with it the population, so the Maya might seem like an odd choice to illustrate sustainable farming practices.[10] But a close study tells instead a tale of extraordinary soil stewardship that sustained the rich Maya culture and its people for thousands of years. Their sophisticated agricultural system has much to teach farmers worldwide.

Some historians have condemned Maya agriculture as classic slash and burn, which involves clearing forests with fire, planting crops for a few years, depleting the limited fertility deposited as ash and charcoal, and abandoning the land. Those historians were greatly mistaken about Maya agriculture. Study of their farming system led to renaming it swidden agriculture, because it uses cool fires that produce biochar, similar to the process that led to the Amazon Dark Earths. Biochar from the Maya forest burns provides soil nourishment, whereas slash and burn typically uses hotter fires, which volatilize organic matter rather than storing it in the soil. In fact, few cultures take such a long view of soil health—the Maya developed a complex agricultural sequence, known as the milpa system or milpa forest gardens, that follows ten- to twenty-five-year cycles. The typical milpa garden intersperses corn, squash, and beans or any of the other seventy food species the Maya use in the rotation. The plant diversity nourishes the soil and provides a rich diet for people. After four years, when crop growth reduces soil fertility and perennial plants become dominant, the Maya cultivate a mixture of tree and shrub species that provide nuts, fruits, cacao, medicinal plants, and building materials. The reforested patch of land is maintained for about twenty years before repeating the cycle. The combination of trees and shrubs creates a balanced ecosystem. Trees capture copious carbon through photosynthesis, replenishing soil organic matter, and retrieve leached nutrients from soil depths inaccessible to smaller plant species. When the trees burn, they once again deposit nutrients in charcoal and ash in the upper horizons that are made available for the next cropping sequence.[11] With several gardens managed asynchronously and simultaneously, there are always some producing the staple crops, such as maize, while others are maintained in a restorative phase, producing other necessary forest products.

How ironic that historians have dismissed Maya agriculture as soil-destroying, whereas in reality, the rigorous planting sequence in the milpa gardens is an example of a highly effective regime to *maintain* soil health. Many historians cling to the storyline that soil erosion caused the Maya decline, even to the point of arguing that contemporary Maya are repeating the sins of a thousand years ago using as evidence Landsat 4 satellite images that show deforestation in regions the Maya inhabit. Yet those same images are known to be in locations where cattle pastures replaced forest. The flaw with their argument is that the Maya did not pasture animals one thousand years ago! Instead, they hunted wild mammals and birds in the forest. Today's deforestation does not provide relevant evidence about the purported Maya collapse.[12]

Many Maya-cultivated tree species persist today in the remnants of their forests, which are as much a Maya legacy as their temples and monuments.[13] The Maya choose trees for the overstory and plants for the understory based on utility and rapid growth, which helps to crowd out weeds. The system works superbly, but it is labor-intensive—requiring seed collection, hand weed management, and burning. Successful milpa gardens rely on knowledge, skill, and labor. Today many farmers lack the labor and time to invest in hand weeding and reforesting gardens, so they resort to labor-saving practices such as herbicide application, which has been accompanied by a loss of almost 20 centimeters of soil in just four years, stripping most of the topsoil. When nutrients are not replenished by two decades of forest growth, these sites become infertile.[14]

Lake Salpetén in Guatemala contains clay sediments that soil scientists have interpreted as indicating three waves of soil erosion during the period of Maya prominence in Mesoamerica. The first two coincided with bursts of expansion of land clearing for new milpa forest gardens. When the forests were burned on virgin land, there may have been immediate soil loss, but the land gradually recovered as the meticulously managed milpa gardens rebuilt a thick soil layer.[15] For almost three thousand years, traditional milpa sustained a growing population on its rich soil substrate. Today's Maya steward soil carefully and demonstrate the beauty of the milpa gardens for growing food in harmony with soil and the broader ecosystem, so it is difficult to reconcile 7 meters of lake sediment derived from agricultural soil with the care and attention the Maya pay to their soil.

The sophisticated soil stewardship of the early Maya adhered to the two principles of erosion prevention: managing wind and water as well as the soil itself. Most of their soil management was embodied in the milpa garden design, which provided plant diversity, good land cover, protection of soil structure with planting sticks rather than plows, and rotation inherent in the garden cycle. In the limited land area where they encountered swampy lowlands they built canals and raised beds to protect the land from flooding, and on slopes they installed terraces to reduce the likelihood of gravity and water washing soil away. Modern remote-sensing technology, which penetrates the dense forest cover that had prevented previous detection, revealed intricate networks of terraces and drainage systems, relics of Maya innovation.[16] Even today, terraces built over one thousand years ago hold three to four times as much soil as nearby unterraced land. To satisfy the second principle of erosion prevention, Maya farmers work to enrich their soil, boosting carbon by deploying deep-rooted perennials, compost, and copious biochar. These practices, combined with the absence of plowing, built and maintained robust soil structure that countered erosive forces.

Today in the Lacandon rain forest near the Mexico-Guatemala

border, fewer than two thousand Lacandon Maya preserve the language and culture. They live among five hundred thousand indigenous people, mainly of Maya descent, continuing to raise food in traditional milpa gardens with a great diversity of plant species and long periods of soil regeneration. The Lacandones who survived both the Maya population decline and subsequent brutal colonization by European invaders demonstrate the sustainability of traditional Maya farming methods, but the community continues to battle for its existence. Intruders abuse the land by clearing forests, monocropping, and introducing cattle grazing that together destroy both the rain forest and the soil. The indigenous group is also under siege from the Mexican government, which offers monetary incentives to deter the Lacandones from burning forests in the milpa cycle, based on a misguided assumption that all burning is destructive. Ironically, without managed burns, the forests are vulnerable to spontaneous hot fires that release carbon into the atmosphere rather than locking it within the soil.[17]

These successful farmers of the Americas offer lessons in soil management that have sustained their crop production for thousands of years. Using an intensive landscape-management scheme, they kept the forests healthy by frequent regeneration, built soil health and aggregation, and simultaneously produced a myriad of food types. This raises the question: Is it imaginable that swidden agriculture could be adopted more broadly? Could today's conventional and industrial farms integrate intensive management of gardens into a cropping strategy that maintains both soil health and satisfactory yields? These interventions may be too radical or difficult to scale up for widespread adoption, but we need to consider every proven practice for enhancing soil if we expect food production to continue at the scale needed to feed the growing global population.

As desertification expands the reaches of arid land, it is essential to learn how indigenous people have farmed drylands for centuries. The Zuni Pueblo Indians in the southwestern United States practice one of the oldest agricultural systems in North America, where they have farmed continuously and combated erosion for millennia.[18] Linguists believe that the Zuni language, unique among Indian languages, may have been used in the same geographic area for seven thousand years. But without written records, language does not leave behind physical artifacts. Fortunately, farming does. Archaeologists have found evidence that the Zuni have farmed the same land continuously for three thousand to four thousand years, indicative of a society that protects its soil.

Zuni ancestors, known as *awu:wu:na:awe:kwi:kowa,* migrated from today's Central America and Mexico about four thousand years ago, bringing with them foods that remain staples of the Zuni diet today—maize, beans, and squash. These plants have provided the primary nutrition for centuries, including during a sharp population expansion that occurred sometime around 1200 CE. At first glance, it is remarkable that anything grows in the arid Zuni land, but despite its barren appearance, this desert is in one of the five most biodiverse regions of North America, supporting organisms that have navigated the challenges of desert life, as have the Zuni.[19] Their farming practices use water judiciously and manage erosion in a climate of extremes.

Appropriation of land by European settlers and the U.S. government shrank Zuni territory from a sprawling 6 million hectares (almost the size of Ireland) to its current size of less than 200,000 hectares (smaller than the city of Johannesburg, South Africa). Despite facing both a harsh environment and invaders who stole the land that contains their heritage, the Zuni have maintained food production using their own innovative farming practices and have protected their soil from devastating erosion.[20]

Zuni lands are largely semiarid Aridisols with some extending onto Mollisols and Alfisols at higher, more humid elevations on the eastern edge of the region. The land is subject to prolonged droughts punctuated with torrential rains that wash away the fragile soil. Erosion is apparent in the arroyos—deep incisions in the land cut by rain. For unknown reasons, extant arroyos first cut the earth between 1880 and 1919 and continued to expand, some reaching 30 meters deep and 50 meters wide. The gushing water that gouges these deep canyons often arrives in a few torrential thunderstorms during July, August, and September.

Episodic access to water creates challenges for soil and crop management that the Zuni handle with runoff agriculture, an approach used throughout the world's arid lands, which involves collecting runoff water from upper, often forested, slopes during the wet season. The Zuni must use elevation to balance temperature and precipitation, inversely related variables, by strategically placing their fields above the valley floor but beneath a hillside that sheds water during torrential rains. The fields capture sediment and nutrients from the higher altitudes while avoiding the floods and frosts typical of the lowest points in the valley. To slow water movement, the Zuni construct small permeable dams of stone or brush that distribute the water and sediments. The dams act similarly to terraces, enabling the water to sink into the soil rather than flow over it. The brush is woven into networks that create a herringbone pattern, engineering channels that redirect water flow. The brush is particularly effective because of its hydraulic properties and semipermeable nature. Over time sediment and nutrients accumulate in the field, creating soil that is far deeper and richer than the surrounding land. To extract further benefit from the limited water supply, the Zuni have bred plant varieties whose seeds can be planted 30 centimeters below the soil's hot surface, where water is available. Seedlings of these hearty plants must grow a long way to reach sunlight! Zuni breeding has also produced maize cultivars that, compared with cultivars in general use in the Southwest, carry more mycorrhizae, the root-infecting fungi that enhance root surface area. The Zuni plants are consequently more efficient at water and nutrient uptake.[21]

Agriculture and land managers have much to learn from the Zuni. They recognized long ago that the desert was not a wasteland, as many European settlers assumed, but instead a land of diversity. White settlers and the U.S. government arrived believing that they needed to teach the Zuni how to farm, but by the 1930s began to realize that the ingenious Zuni methods for water management were far more successful than anything introduced from outside the community for erosion control.[22] By listening to the land, learning its cycles, and devising management practices to prevent soil movement and encourage its resilience, the Zuni have coaxed an abundance of food from the desert, nurturing the soil while nourishing themselves. Many farmers can draw on wise Zuni stewardship, and the philosophy underpinning it, to build soil legacies that will persist for centuries to come.

The Māori people of Aotearoa, or New Zealand, have protected soil through practices shaped by their deeply spiritual and cultural connections with the land. Their fight to preserve soil for centuries has placed them, like many indigenous peoples around the world, in conflict with European colonizers. The Māori endured seeing their land extensively deforested, degraded, and governed by a colonial ownership and management system. They have adapted to an imposed legal system in order to protect and rehabilitate the land using both traditional and conventional management styles, combining spirituality and science to navigate a system that is not their own.

They offer lessons in the power of human resilience in building—and rebuilding—land legacies.

Unlike the indigenous people of the Americas, who speak many languages across hundreds of nations that sprawl over a large continent, Māori share a common language and live mostly within New Zealand's two, relatively small, main islands. They arrived in New Zealand from Polynesia around 1350 CE and lived as hunter-gatherers before forming settlements surrounded by crops and horticulture. Their land management was based on knowledge and skills originating from Polynesia with a rich spiritual connection to the land and broader environment. In fact, Māori are known as *tangata whenua*, people of the land, illustrating their belief that humans do not own land but belong to the land and should give back to it what they take.[23] This mutually giving relationship is expressed through the common tradition of burying placentas to cement each child's bond with Papatūānuku, Mother Earth, and is linguistically embodied in the word *whenua,* which translates as both "placenta" and "land."

Today Māori rights and land guardianship remain precarious. Large tracts of land were lost mainly during the nineteenth century under a raft of colonial laws, and now Māori retain only 6 percent of their original traditional land area, despite filing hundreds of land claims and grievances. Their land is split into thousands of fragmented blocks, each with multiple landowners and a separate governance entity based on Māori land legislation passed in 1993. About 85 percent of Māori live in cities, leaving small rural communities to maintain the tribal homelands, or as they say, *ahi kaa,* which means "to keep the home fires burning." But both urban and rural Māori draw on traditional knowledge to develop new soil management practices that nourish their relationship to tribal areas and Māori land. Their spiritual and cultural connection with all living things impel the

Māori to become students of the soil, developing their own soil terms and classifications. They assess soil health largely through their senses, noting its color, smell, feel, and texture, and they use the term *mauri* to discuss the vitality of a soil and its ability to support diverse life-forms and to ensure well-being.

Confronting economic, social, and political forces that take and degrade the soil, Māori have sought to protect their traditions and beliefs about soil, especially their view of it as a living system that gives the people a sense of identity and unity.[24] Māori environmental researchers Jessica Hutchings, Garth Harmsworth, and colleagues have created a new framework for soil health based on Māori values of ancestral lineage, interconnectedness, and divinity. The framework reinforces Māori rights of sovereignty and guides them to fulfill their guardianship responsibilities for land to improve its *mana,* a word that has many layers of meaning, including power, honor, prestige, fertility, and health. Māori farming and cropping practices elevate a soil's mana by enriching its fertility, increasing microbial diversity with nitrogen-fixing plants, adding compost and other amendments, and avoiding chemical additives and mechanical degradation.

In the nineteenth century, British settlers sought to profit from New Zealand's natural resources and brought a starkly contrasting, nonspiritual view of soil, land, and ownership to New Zealand. Despite much resistance by Māori, the British convinced hundreds of chiefs to sign the Treaty of Waitangi in 1840, a document that was interpreted very differently by the Māori and British and remains the subject of ongoing debate. The treaty abruptly resulted in colonial appropriation of more than one million hectares of land and infrastructural projects that scarred Māori landscapes with deforestation, sprawling pastoral agriculture, roads, drainage systems, and British settlements.

To a large extent, Māori have now accepted the treaty as a constitutional basis of Aotearoa–New Zealand, with the Māori version of the treaty providing a high level of indigenous rights and participation. New treaty legislation has also helped to elevate rights for Māori in tribal areas and across New Zealand, supporting them to take a more prominent role in decision-making over national natural resources. Internationally, this model of co-management and co-governance drew attention when Māori concepts infiltrated mainstream legislation and policy with historic legislation in 2014 and 2017 that awarded personhood to a tract of conservation land and a river.[25] Many Māori hope that soil will be next to gain the recognition of personhood.

New Zealand's Waiapu catchment in the eastern North Island is an area of steep hills covered in fragile soil. It is the spiritual and cultural home of the Ngāti Porou tribe. After thousands of years of stability under native forest, the catchment is now among the worst erosion sites in the world with very high sedimentation in the rivers and streams. While under the native forest canopy and leaf litter, the soil was protected from the high intensity rainstorms and tropical cyclones that frequent the area. Traditional Māori agriculture and natural forest management had also enriched the soil with a compost made of shells, seaweed, and discarded plant material. After colonization, deforestation, and a transition to pastoralism, the catchment became highly degraded and lost thousands of tonnes of soil. Without its protective covering, the sloped land became pocked with deep gullies, and sediments slid into the Waiapu, a river of profound spiritual significance to the Māori. Today soil erodes at a staggering 180 tonnes per hectare per year, muddying the sacred river, raising its bottom by 2.4 meters per year, and damaging local floodplains and Māori farms (see pl. 3).[26]

Like so many other places in the world, New Zealand today is experiencing the effects of climate change in more frequent, heavy rainstorms that pummel the land with high-velocity raindrops, driving soil off hillslopes at an alarming rate. What's more, almost 60 percent of New Zealand is classified as mountainous, hilly, and vulnerable to hill-country erosion. Reforestation efforts across New Zealand focus on creating a naturally diverse forest that will protect, rebuild, and anchor soil in place. Māori have been at the forefront of these soil conservation and land development initiatives. One tree species that has attracted special interest because of its cultural and economic value to Māori is the mānuka (*Leptospermum scoparium*), which is native to New Zealand and provides nectar for pollinating western honey bees. Mānuka trees are also very important to soil conservation in the east coast region that includes the Waiapu catchment. The bees themselves have a remarkable history, having been imported in the late nineteenth century by a British beekeeper, Mary Bunby, who kept them alive for the six-month voyage from England to New Zealand. She made the first honey on the island, probably feeding the honey bees on mānuka nectar, a practice that has since expanded. Māori medicine (*rongoā*) employs the honey for healing, and in the twenty-first century mānuka honey reached Western shores, where it has taken hold, touted by biochemists, health commentators, and actors for its antibacterial and wound-healing properties.[27] Today the bees are helping reforest and protect Māori land, reducing erosion, and increasing the income of the indigenous community. The mānuka tree may be the most famous and lucrative example, but farmers and foresters have introduced other species into the Waiapu catchment to protect the soil and restore the once lush ecosystem.

Planting trees to restore the landscape is an effective strategy to protect and build soil, and in many locations, it is the best way to do so. Tree roots are terrific soil managers, binding soil particles, feeding microorganisms, and recruiting nutrients from deep in the soil profile to the top horizons. New Zealand's government aims to plant trees throughout eroded areas, but misguided policies have alienated many Māori. The government's Emissions Trading Scheme, for instance, provides incentives for reducing atmospheric carbon by planting exotic trees such as pines, whereas Māori yearn to restore native forests from their *whakapapa* (ancestral lineage) with the diversity that builds soil mana and supports the ecosystem. In the nineteenth century the Europeans defiled Māori land by replacing forests with pastures; today the government repeats that insult by designing solutions that disregard Māori spirituality, traditional practices, and knowledge of the land. Although individuals are no doubt attempting to bridge the gulf, the government as a whole stands in stark contrast with Māori philosophy and land practices.

Māori are beginning to take a stronger role in developing New Zealand's land restoration policy, applying their knowledge of wise stewardship. They hope to partner with the national government, local government agencies, industries, and a range of other stakeholders to reclaim their rights and the ground of their cultural heritage. The rest of the world can draw inspiration from the Māori's spiritual and cultural attitude toward land. Their partnership with the soil—indeed, their belief that they are *of* the soil—taught them to be talented land stewards. Even cultures lacking such a spiritual link with the soil can learn from the outcomes of that spirituality.

As we study the underground ribbon that connects the past and future, what can we learn from indigenous soil legacies? First, plant diversity is a striking feature common to several, such as the Maya milpa gardens, Zuni runoff agriculture, and Māori forests. Each system capitalizes on many local species to nourish and anchor the soil and support a robust ecosystem. Plant diversity is the antithesis of conventional modern agriculture in which monocropping rules the field, weeds are eradicated, and, depending on commodity prices and biofuel markets, farmers may plant the same crop for several consecutive years. The second commonality is water management. The Maya and Ifugaos use terraces, and the Zuni build dams to reduce the velocity of water and to redirect it. The Māori depend on trees to intercept raindrops, reducing their impact on soil aggregates when they land. The third theme is the application of organic amendments to the soil. In Papa Stour, the farmers enriched the soil with manure, turf, and seaweed, producing plaggen soil. The Maya and Amazonians used weeds and other plant waste, the Māori use seashells, fire ash, and seaweeds, and the chinampas are stuffed with peat. All of these practices produce black soils that are richer than the surrounding earth in nonagricultural sites. And finally, none of these land stewards plow their fields. All have minimal impact on their soils through the use of sticks, spades, or hand planting, enabling the soil to build robust structure. The lessons are reinforced by the failures. When the Maya ceased the twenty-year soil regeneration cycle, their milpa system struggled. When the European immigrants deforested Māori land, New Zealand developed some of the worst erosion sites in the world.

History is full of agricultural systems that have failed and disappeared, often along with the societies that depended upon them for sustenance. Others persist after centuries or millennia of continuous practice, stitching the ribbon that links the past and future. Soil that is nurtured and protected grows deep and dark, leaving an inheritance of both soil and the blueprint to make more of it. Most agricultural practices are on track to deplete much of the Earth's soil in a few short decades. Looking at the cultures that have sustained their soil throughout most of agrarian history, we should be appalled at the current trajectory of conventional agriculture. We can do so much better. The principles are well known, and the practices are adaptable. The burden of protecting soil cannot be relegated to indigenous people and environmental activists. All of us who depend on mainstream agricultural production need to demand wholesale change in management if we plan to continue to eat.

9

Soil Heroes

When I was a child, one of my favorite picture books was about a farm. Its illustrations depicting rolling hills verdant with crops and dotted with cows, a farmer on his red tractor, and a girl carrying pails of fresh milk from the barn to a tidy white house surrounded by colorful flowers portrayed a life of bucolic bliss. A life that looked peaceful, enduring, and safe. Modern agriculture across the world is quite distant from this romanticized image. It's a demanding life full of tough choices and uncertainty. Many people farm to produce food for their own families. Others inherit land with generations of toil and history buried in it and feel obligated to continue farming. And some swear by the lifestyle and will make sacrifices to protect their agricultural livelihood. But none of these farmers have easy lives.

As an adult I learned the truth behind my picture book, and farmers became my heroes. There are few jobs as complex as farming. Consider the many skills farmers need and how many ways the landscape—literal and figurative—is always changing. They need deep knowledge of the nutrition, diseases, and life cycles of the plants or animals they raise in order to match the right cultivars, fertilizers, pesticides, and feed. They must optimize planting, tillage, harvest, and animal management regimens for the local weather, accounting for both long-term trends and today's surprises. The constancy of change means that every decision needs to be revisited each year.

On large, mechanized farms, farmers must fix intricate equipment, from milking machines to tractors, and navigate accounting, global and local markets, interest rates, and loans. They need to adhere to the regulations of their country and state or province with an eye on those of other countries where their goods will be marketed. Regulations and food preferences change locally and globally, altering the desirability and profitability of certain commodities. Those who sell their goods in national or international markets face unpredictable supply and demand. Sudden weather changes or new pests and pathogens that destroy crops or livestock can transform commodity markets overnight. Farmers must make choices for an entire season before weather and markets reveal themselves.

The demand for particular foods skyrockets or plummets when a new health claim appears or an old one is discredited. Consider the almond craze that was prompted by publications espousing the nut's health benefits. In the United States alone almond demand has increased 7.5 percent every year from 2015 to 2020, and similar growth is projected through 2028 worldwide.[1] But almond trees take five to twelve years to begin producing nuts, making agile responses biologically impossible. Demand may subside as suddenly as it ballooned, saddling farmers with unrequited investments. Farmers need a bit of pluck and a pinch of clairvoyance added to all their practical skills.

The skill, risk, and investment intrinsic to all farming make it unsurprising that many farms barely survive, but smallholders face a barrage of unique challenges. Three billion people live on approximately half a million farms smaller than 2 hectares, which produce the majority of food in many countries in Africa, Asia, and Latin America. Despite their essential role in feeding the population, more than half of these smallholders live below the poverty line in many countries, including Malawi, Vietnam, Bolivia, and Guatemala.[2] Many lack the capital to invest in seed, fertilizer, or machinery, preventing them from increasing their yields or reducing labor needs. Smallholders are also liable to bear the extra challenge of balancing the uses for the food they produce. The same harvest may generate income, meet their families' immediate dietary needs, or serve as seed for planting the following season. For smallholders who lack reserves, the consequences of these decisions can be profound.

Many large farms have similarly marginal profits, leaving farmers vulnerable to the unexpected, and the first twenty years of this century have delivered the unexpected in abundance. In 2001, the foot and mouth epidemic in the United Kingdom led the government to require livestock managers to slaughter all susceptible animals, causing catastrophic damage to the meat industry. A long recovery followed. The COVID-19 epidemic dealt blows to farmers worldwide, causing a global decline in meat, dairy, and biofuel prices. Some operations struggled with labor shortages caused by immigration restrictions.[3] Just as the coronavirus arrived, locusts caused the worst agricultural losses in decades, leaving many farmers on the brink of economic catastrophe. The year 2020 was a menace.

The constancy of economic insecurity can take a deep psychological toll on farmers. Worldwide, suicide is more common among farmers than in almost any other profession. A 2014 report documented that farmer suicide increased fivefold between 1997 and 2012 in Punjab, India, and was 50–100 percent higher than in the general Indian population. In 2009, a farmer committed suicide in India every thirty minutes. In 2016, the Centers for Disease Control and Prevention reported that the suicide rate among U.S. farmers was three times higher than in the general population and identified agriculture as one of the top five sectors of highest concern for male suicide.[4]

The risks of farming have contributed to the trend toward larger farms, which turn higher profits than smaller ones.[5] Large farms breed industrial farming, which is characterized by high use of monoculture, fertilizers, pesticides, antibiotics, and irrigation water. Industrial farming causes loss of biodiversity, dried-up aquifers, and soil degradation. Large industrial farms are the result of consolidation of the midsized farms that used to be the backbone of agriculture in the United States and many other countries, from Mongolia to Brazil, where industrial farming has taken root. Many people assume that industrial farming has replaced family farms—defined as those on which the principal operator and his or her relatives own more than half of the farm business. In fact, today family farms represent 96 percent of U.S. crop-producing farms. Even among extremely large farms, the most likely to use industrial practices, 86 percent are family operations, indicating that type of ownership and farming methods are not necessarily linked.[6] But whether owned by families or corporations, industrial farming is stripping the land of its resources and squeezing out midsized operations.

This is the backdrop for soil stewardship decisions. The context is essential to understanding why implementing good soil practices happens despite the odds, and it must inform the design of new policies that enable widespread stewardship. Financial survival necessarily drives individual choices. Despite their commitment to the health of their land, many farmers cannot afford the risk entailed in making changes. Although increased profit may accompany adoption of soil-friendly practices in the long run, initial costs can be prohibitive. Policy must reduce the barriers to farmers' adoption of new practices.

Conservation agriculture is a sustainable approach to farming that prioritizes soil quality. Among its practices are using trees to reduce wind speed, planting cover crops to protect the soil between main crops, using compost to boost organic matter, and intercropping to fortify soil structure and prevent its migration. No-till planting reduces plowing, thereby preserving soil structure. Canals and pipes reroute water away from fields to prevent flooding and soil loss. Farmers limit the effect of gravity on soil movement with contour plowing around hills rather than up and down them—horizontal plowing creates ridges that impede water and soil movement, whereas vertical plowing creates channels that offer a convenient route for soil to slide away. These are the soil conservation practices available to today's farmers.

Conservation agriculture developed in the United States driven in part by the Dust Bowl of the 1930s, which devastated the plains of Oklahoma and surrounding states. The same farming practices that increased vulnerability of the plains to drought and wind erosion had made other parts of the country susceptible to water erosion. In the northern Midwest, many farms had lost all their soil within twenty years of first plowing. By the 1930s many were desperate for solutions when Aldo Leopold, often called the father of the American conservation movement, lobbied President Franklin Roosevelt for a soil conservation demonstration project. In 1933 President Roosevelt's newly formed U.S. Soil Erosion Service selected Coon Valley, Wisconsin, to host the first soil conservation project in the United States. Coon

Valley is a watershed that had been devastated by nineteenth-century farming on its steep hills, where erosion had cut gullies so deep that they could no longer be farmed or even crossed. Coon Creek was subject to frequent flooding and became sedimented with farm soil that made the water shallow and warm, killing most of the local trout. Leopold and his colleagues collaborated with farmers to rebuild the valley using a suite of methods that remain at the core of conservation agriculture. Contour farming, replanting woodlands, terracing steep land, and planting strips of deep-rooted legumes amid the corn healed the Coon Valley soil (see pl. 6, bottom). Erosion eventually fell by 75 percent and sedimentation of the creek by 98 percent. Farmers experienced unquestionable financial benefits—between 1934 and 1942 their incomes rose by an average of 25 percent.[7]

Despite this and subsequent successes, not all U.S. farmers followed conservation practices, and even those who did still plowed the soil for planting, so it continued to erode. Then, forty years after Leopold's Coon Valley project, a dramatic new practice called no-till planting appeared, which eliminated plowing at planting time and obviated the need for plowing after harvest. In contrast to conventional cropping methods, no-till preserves crop residue from the previous crop by drilling seeds directly into the soil through the stubble rather than dropping them into a plowed furrow. Those were heady days in agronomy. The year I graduated from college, the first study of no-till methods was published, reporting a dramatic 75 percent reduction in soil erosion compared with conventional methods such as moldboard plowing.[8]

Across the world, average soil loss on no-till land is no greater than on nearby nonagricultural lands with native vegetation. By contrast, tillage increases erosion on average ten- to one hundredfold. Over the past few decades, copious research has demonstrated that without tillage, crop yields are as high or higher than with conventional techniques, water use is more efficient, energy use is 7–18 percent lower, and carbon emissions are reduced by two-thirds. Just after no-till was introduced, optimistic researchers estimated that by 2010, 78 percent of the major crops in the United States would be grown with no-till farming methods. Alas, the only countries to approach this goal are in South America, including Brazil, where 74 percent of cropland is under conservation management, and Argentina, where conservation methods are used on almost all the cropped land. In contrast, the United States currently manages only about one-third of its cropped land with no-till practices, and worldwide, only 13 percent of all agricultural land is under no-till management. Uptake of the methods was slow until the 1990s. Between 1999 and 2013, conservation agriculture expanded an average of 8 million hectares, or only 0.5 percent of global farmland, per year, which masks the good news that during the second decade of this century many countries began rapid expansion of conservation agriculture. Between 2009 and 2013, for instance, the amount of China's land under conservation management increased sixfold, Ukraine's sevenfold, and Mozambique's an impressive seventeenfold. During the same period, countries such as Syria and Turkey, which began with no agricultural land under conservation management, adopted the methods widely; by 2015, Syria and Turkey had converted 30 million and 45 million hectares, respectively, to conservation agriculture.[9]

Despite the damage to soil, farmers continue to pull plows across their land because of the benefits of tillage, primarily for weed control. Tillage macerates weeds that are growing and buries seeds of those that are dormant. Farmers who transition to no-till practices must adopt alternative weed-control measures such as crop rotation, fallow periods, and herbicides. Two types of chemicals combat weeds: nonselective herbicides, which kill all plants, and selective ones, which kill weeds and *not* the crop plant. Atrazine is one of the most frequently used selective agents. When it was introduced in 1958, many agronomists and farmers hailed atrazine's ability to kill a broad spectrum of weeds while leaving plants such as corn, sorghum, and sugarcane unharmed. Today atrazine's broad adoption raises concerns because it is mobile in the environment and toxic to animals. It is one of the pesticides most frequently detected in groundwater, streams, and rivers, where it induces mutations and disrupts the animal endocrine system, altering sexual development and causing deformities in amphibians.[10] Farmers have difficult choices weighing the risks to their crops, the soil, and the environment. But sometimes we aren't even having the right conversation about risk.

An entirely new approach to weed control was introduced in 1996 with the marketing of Roundup Ready soybeans. These plants were constructed by transferring a bacterial gene for resistance to glyphosate, the active ingredient in the herbicide Roundup, into crop plants, creating a genetically modified, or GM, plant. The consequence was the conversion of a relatively safe, nonselective herbicide into a highly selective one that kills most plants while leaving the resistant crop plants unharmed. The company profits from the Roundup Ready cultivars when farmers purchase the seed and again when they buy the herbicide. The patents on the plants require farmers to buy new seed every year rather than saving seed from one year's harvest for planting the following year. Some studies show that farmers have realized higher profits from glyphosate-resistant soybeans and cotton, but these profits are dwarfed by the income they generate for the seed and herbicide manufacturers.[11]

We don't know the full extent or nature of the effects of blanketing the land with herbicide-resistant plants, which represent 60 percent of the world's soybeans and almost all the corn and soybeans planted in the United States.[12] There is no precedent for such widespread deployment of a gene that is new to the plant species accompanied by a fifteenfold increase in glyphosate use in just two decades. Farmers and scientists have raised concerns about the genetic uniformity in cultivars, potentially creating a widespread vulnerability in our major crops. Researchers have documented the emergence and spread of glyphosate-resistant weeds following introduction of the resistant crops.[13] What other consequences have not yet been detected? Unfortunately, these documented risks have been obscured by evidence that is still controversial about a link between glyphosate exposure and certain types of human cancers and the vague argument that engineered plants are not natural.[14]

The concern that genetically engineered plants are not natural is particularly problematic. If "natural" is defined as occurring in nature without human intervention, then little about agriculture is natural. Modern crop cultivars have been highly bred, making them quite distant from their wild ancestors. Plants grown in monoculture with synthetic fertilizers and pesticides are definitely not natural. And massive steel blades raking through the soil are not natural. A far more productive dialogue would focus on the potential and measurable impacts of genetically engineered plants on people and the environment, particularly compared to other agricultural practices.

For all the hullabaloo about herbicide-resistant crops, a point lost from the public discussion is their benefit to soil. They eliminate the need for tillage to control weeds, leading to better soil architecture and less erosion. Yet several countries ban GM foods, and they must be labeled in the European Union and other members of the global marketplace, deterring some farmers from planting GM crops. Organic-certified food in the United States cannot contain GM plants, leading many organic farmers to use intensive tillage to control weeds, thereby weakening their soil. The balance of risks and benefits of GM plants remains murky, but two things are clear: soil health belongs in the discussion, and scientists and engineers owe farmers alternatives to tilling and herbicides for controlling weeds.

Weed control illustrates the complexity of farmers' choices. In addition to considering the impact of tilling on their soil, they may weigh the proven animal toxicity of atrazine, the possible carcinogenicity of glyphosate, the known and unknown environmental effects of herbicides and genetically engineered plants, and whether they are targeting markets that ban GM plants. All of these choices are made in the context of their farm's idiosyncrasies and the need to make sufficient profit to survive another season.

Cover crops offer another means to protect soil. Leopold was among those who advocated use of cover crops to prevent erosion between harvest and the next planting. Although autumn plowing is favored by many farmers to subdue weeds and enable earlier planting the following spring, it ruins structure and bares the soil to the caprices of the weather for up to eight months. Prolonged exposure to wind and water is a crime against the land. Climate change amplifies the problem in many northern parts of the world because snowfall is declining, leaving the land without a winter blanket. Cover crops protect the soil and support the microbes that continue to work busily during the fallow winter months, encouraging an active, abundant microbial community come spring.

Dozens of plant species can serve as cover crops, including legumes (such as clover and vetch), mustards (such as canola and daikon radish), and small grains (such as rye, quinoa, and oats). The plants typically grow sufficient roots and leaves before the frost settles, intercepting wind and water during the winter, providing a bulwark against soil movement. In the spring, the cover crop can be harvested, killed with chemical treatment, or plowed into the soil to enhance its nutrient status. Seeds of the main crop can be drilled directly into the cover crop, letting it continue its task of protecting and nourishing soil and its microbiome.[15]

Rotation of crops also replenishes soil. A nutrient depleted by one plant might be replaced by another, and depriving hungry microbes of their best host for a period of time interrupts pathogen cycles. Although most farmers know the benefits of rotation, many grow corn continuously, year after year, if the market price is high because corn is more profitable than most alternative row crops, particularly with incentives to grow corn for biofuel production. Among modern crops, corn is also the most destructive to soil because its puny root system leaves little residue behind. Continuous planting of corn under conventional management degrades soil by sucking out its nutrients, damaging its architecture, and reducing its resilience. There is a saying among soil scientists that for every kilo of corn harvested, the field loses a kilo of soil. The world produces 1 billion tonnes of corn, so if the saying is true, conventional corn production would be responsible for a substantial amount of the 24 billion tonnes of soil lost each year. Rotating corn with soil-nourishing crops is essential to maintaining soil nutrients, organic matter, structure, and moisture-holding capacity. In temperate regions farmers often rotate corn with soybeans. In dairy regions, they may include alfalfa in the sequence, which they feed to dairy cows as hay or silage. Soybeans and alfalfa are both legumes, and thus their nitrogen-fixing bacteria add nitrogen to the soil, and both have robust root systems. Yet despite the demonstrated value of crop rotations in building soil health and reducing erosion, the number of midwestern farms in the United States planting corn on the same land for a *fourth consecutive year* doubled in 2013. Profits are necessarily the dominant driver for farmers planning to farm another year.[16]

Intercropping offers an alternative or addition to crop rotation.

Unlike the Maya intercropping, in which the plant species were mixed together, large-scale intercropping typically involves replacing only a small portion of the main crop with deep-rooted prairie plants that slow or prevent water movement across the land. Strategically placed prairie plants reduce erosion by 95 percent and have side benefits, such as providing pollinators with a habitat and reducing emissions of nitrous oxide from the soil.[17] Incontrovertible evidence demonstrates the salubrious effects of reduced tillage, cover crops, crop rotation, and intercropping, but they remain underused.

Conservation methods clash with the reality of farming when they are costly to farmers. Cover crops require extra seed. No-till planting requires new equipment. Crop rotation requires planting less profitable crops for several years to replenish the soil. Intercropping corn with deep-rooted perennials reduces the amount of land a farmer dedicates to the main crop. In the United States, intercropping reduces crop insurance, which is based on the amount of land that will be planted with the main crop compared with the previous season: 10 percent less land in corn means an intolerable 10 percent reduction in crop insurance. In the long term, most soil conservation methods pay off financially because they increase fertility and lower the incidence of crop disease, thereby diminishing both fertilizer and pesticide costs. In Brazil, the benefits of no-till practices have boosted land value by 50 percent, illustrating the long-term financial gains.[18] Soil-protective practices have vast benefits that outweigh the modest costs, but even a small initial financial toll prevents their adoption in fragile agricultural operations with narrow profit margins.

Today agricultural movements such as permaculture, organic farming, and regenerative grazing have taken hold to different extents. Permaculture, which most closely approximates the Maya milpa system, began in Australia in the mid-1970s with a belief that gardening is the most sustainable form of agriculture. Its practitioners subscribe to principles of local production and consumption, understanding the land and flow of water through it, and minimizing waste and energy use. These principles led to food production systems that incorporate biologically diverse plant communities. Permaculture practitioners observe their land's topology and water flow to design areas for crop production and water collection. They rejuvenate soil by building up organic matter and minimizing disturbance by practices such as tillage. In 2021, the Permaculture Worldwide Network listed 2,655 permaculture projects worldwide, including some managed by Maya farmers in Mexico.[19]

Organic agriculture offers another alternative to typical modern farming methods and has been adopted far more widely than permaculture. Its methods are predicated on four principles: health, ecology, fairness, and care.[20] These refer to maintaining the health of the planet, working with the ecosystem and its cycles, ensuring fair use of environmental resources, and protecting the system for future generations. Although these high ideals are a bit nebulous, they have evolved into a set of practices that omit chemical pesticides and fertilizers and protect the soil.

The organic movement began in the early twentieth century, propelled by the work of Sir Albert Howard in India, F. H. King in Wisconsin, and Rudolf Steiner in Germany. These pioneers advocated building soil health with manure, compost, cover crops, and crop rotation and using biological means for pest control. The movement gained momentum in Switzerland and Germany, where the first standards for organic farming were developed between 1928 and 1933. In the United Kingdom, Lady Eve Balfour emerged as a powerful advocate for organic farming in the 1940s and 1950s. She wrote a book that was seminal for the organic food movement, and she cofounded the Soil Association, which advocated sustainable farming methods. She also started the first long-term experiment in organic farming, known as the Haughley Experiment, which revealed that under organic management the soil gained fertility and a higher density and diversity of earthworms. Today, Lady Eve's Soil Association certifies 70 percent of the organic food grown in the United Kingdom and continues to campaign for farming methods that enrich soil and prevent its loss. The early organic movement challenged the utilitarian approach to land, recognizing soil as an ecosystem, a living machine that can be strengthened or weakened. The practices of the movement recognized the importance of nurturing soil to ensure that there would be soil in which to grow the next crop and the one after that.

Organic farming has expanded since Lady Eve Balfour championed it in the mid-twentieth century. In 2018 the worldwide organic food market topped $165 billion, and this is projected to quadruple by 2027. Although only 1.5 percent of farmland globally is dedicated to organic production, it represents a substantial proportion in certain countries—38 percent of Liechtenstein's agricultural land, 34 percent of Samoa's, and 25 percent of Austria's. The land area dedicated to organic farming in the European Union increased almost sevenfold in the first twenty years of this century, a period during which worldwide organic food sales increased fivefold. In the United States, organic production has not kept up with domestic demand, prompting importation of organic foods—75 percent of soybeans, 50 percent of corn, more than half of organic fruit, and one-third of all vegetables. One concern is that the lower average production of organic farming—perhaps as much as 30 percent—may require increasing land dedicated to food production, which would not benefit soil.[21] Productivity, land use, and soil management need to be prominent in the maturation of organic agriculture.

Not all modern farming fits neatly into the categories of conventional (with high tillage and chemical use), permaculture, or organic. Rather than subscribing to an orthodoxy, some farmers adapt methods to their local landscape. That's the way Joe Bragger does it. He farms in the "driftless region" of Wisconsin, where steep hills and deep valleys persisted after the last glaciation event that flattened most of the state. The Bragger farm lies outside the town of Independence on the far western border of the state. For an urbanite, driving through Independence feels like entering a time warp, transporting the visitor back to a less chaotic time, with its Romanesque Revival–style town hall–opera house built in 1908, advertisements for a Friday night Sow Belly BBQ at a local brewery, and a vintage gas station displaying Texaco's emblematic red star and an antique gas pump with a shiny crimson paint job. Despite the twentieth-century appearance of the town, Joe Bragger's farming is as twenty-first century as can be.

The slope of the land in the driftless region alerts any student of soil to the hazards of farming. Relying on the two principles of erosion prevention, Joe constructed dams and waterways to control water flow and capture nutrients before they leave the farm, and he uses cover crops, no-till planting, and crop rotations to enhance carbon and soil structure. He describes the effect of cover crops with the fervor of those who have experienced their power firsthand. One summer, 12 centimeters of rain fell in a single storm, removing several centimeters of soil from a field planted with corn alone and little from the one with a cover crop. Later that season, Joe was surprised to see that the corn in the field with the cover crops stood tall and straight, whereas the corn without a cover crop was short and leaned like that proverbial structure in Pisa. Now he never leaves fields bare, planting rye or another cover crop in the fall. As weather patterns have turned extreme in recent decades, the farm has matured, benefiting from years of conservation management so that it now excels in the violent rainstorms that have become commonplace during Wisconsin summers. A historic storm in June 2013 dumped 35 centimeters of rain in a single day, submerging nearby bridges, roads, and farmland and washing vast amounts of soil into creeks, but the Bragger farm suffered no damage. The soil had accumulated sufficient carbon to maintain its structure and anchor it to the steep hillsides even in the face of an epic deluge.

Joe Bragger doesn't just believe in his farming methods—he has data to show they work. From 2002 to 2008 scientists from the University of Wisconsin Extension Program studied his farm, demonstrating that his farming methods conserve both soil and water. They showed that his no-till fields captured a whopping 98 percent of the water delivered by the 35-centimeter rainstorm! The Bragger farm is fortified against the gathering storm of climate change, but it may not be as well prepared for the coming economic weather. The encroachment of large farms that monopolize markets, as well as low milk prices and fallout from COVID-19, all challenge the family farm's existence. Its conservation farming methods may make it more profitable but may not be enough to ensure its survival without national changes to agriculture's financial incentive system.

Africa has more degraded land than any other continent. With 700 million degraded hectares—an area the size of Australia—it offers an opportunity to make substantive environmental strides through landscape restoration. The African Union made the extraordinary pledge to restore 100 million hectares of degraded land by 2030. Many African countries are exploring trees as their tool of choice for restoring the land and combating climate change. Both agroforestry, in which trees are interplanted with crops, and silvopasture, in which trees are integrated into livestock production, can restore soil carbon and nitrogen, increase crop production, and enhance the livelihoods of millions of people in the region. The diversity of products harvested from these systems increases the resilience of the land and people by reducing vulnerability to crop failure and desertification, restoring the continent's biodiversity, and reducing dependence of families on a single staple food.

A particularly effective form of reforestation is a traditional agricultural practice known as farmer-managed natural regeneration, or FMNR, which creates a mosaic of crops, livestock, and trees. The distinguishing feature of this practice is that the trees are regrown from live stumps already in the soil rather than establishing them from seed. Stumps regenerate trees more quickly and require less labor to start the process (fig. 16).[22]

In Niger, FMNR is credited with increasing grain production by 500,000 tonnes, equivalent to $900 million, which feeds an additional 2.5 million people. In some regions, the proximity of trees to dwellings increases access to firewood, reducing the time women spend collecting wood from 2.5 hours to 30 minutes per day. Many have freed up enough time to start their own businesses. A study of Niger, Burkina Faso, Mali, and Senegal—Sahelian countries that suffered terrible damage from the 1980s drought—reported that agroforestry has boosted household income considerably. Farmer-managed natural regeneration introduced a mixture of trees grown for fruit, lumber, and nitrogen-fixation capacity. Intermingling the trees with cereal and vegetable crops was lucrative. In Niger, for example, households with twelve members increased their income by an average of $72 per year by adopting continuous use of agroforestry, a substantial gain in a country where the household income averages $617 per year. Further benefits of FMNR include variation in people's diets, diversified in-

Figure 16. New growth from a tree stump. Illustration by
Liz Edwards.

come sources, and increased soil fertility. In North Ethiopia, this agricultural practice has revived sixty springs, improved the groundwater recharging potential, diminished runoff and erosion, and increased
food production more than fivefold for 5,130 households. Honey
production has taken off, building materials are more accessible, and
the resilience of both families and the land has increased.[23]

Scientists have been surprised by several FMNR-related findings.

A 2020 study of 316 plots in Ghana and Burkina Faso reported that more degraded land benefited most from trees, suggesting that the regrowth was dominated by species adapted to poor soil conditions. Other research indicates that sandy soils experience the greatest gains in soil carbon.[24]

This method of farming is ecologically and economically sound, and its effects are rapid. Expanding the land under FMNR management could dramatically improve the livelihoods of smallholders in Africa and lower atmospheric carbon by increasing carbon deposition in soil.

Another cutting-edge, promising direction for soil conservation lies in the realm of animal husbandry. Most of the world's farmland is home to livestock, making their habitats vital to preserving the Earth's soil. A curious paradox lies in the history of herds of large animals in the wild and on ranches. Herds of bison, elephants, llamas, and others migrated across savannas and plains for millennia, their waste contributing to creation of some of the most fertile, deepest soils in the world. Yet modern-day ranching is blamed for soil erosion and desertification. A new look at the impact of cattle herds under various management practices suggests an explanation for the paradox.

Advances came from development of regenerative livestock agriculture, which is based on intensive rotational grazing. This practice requires grazing cows in dense herds on small sections of pastures and moving them frequently. Using fencing, the farmers crowd the cows together to eat about half of the plant biomass and then move them to a new area, repeating the process daily or even more frequently. Research shows remarkable results. After switching to regenerative grazing, the Oasis farm in Botswana doubled the size of the herd with no adverse impact on the land. The Estancia Nevada ranch in Chile started with badly eroded land, which recovered rapidly and sprouted new plant species that increased soil carbon. Switching to regenerative grazing enabled a bison ranch in South Dakota to increase the herd size fivefold, which was followed by a doubling of plant biomass, a tripling of water infiltration into soil, and a surge in soil carbon. On another northern U.S. ranch, the soil's water-holding capacity tripled. In Zimbabwe, cattle ranchers are now routinely rejuvenating their land and reversing desertification with regenerative grazing, observing increases in soil carbon and organic matter, more active microbiomes, and greater plant diversity. A study of eighteen conventional and seven regenerative ranches in Chiapas, Mexico, reveals that the regenerative ranches supported more cows per hectare; in addition, adults and calves had lower mortality, and they required fewer chemical inputs. The soil was deeper, more aerated, and more densely vegetated.[25] Initial concern about the welfare of the cattle has been replaced by evidence that the crowded conditions better approximate wild herds and may be less stressful for the cattle.

Conventional beef production is a target of climate concerns because of the volume of methane produced by cows. But some people argue that regenerative grazing makes beef production a net *consumer* of greenhouse gases! One study showed that conventional practices led to the release of 6 kilograms of carbon dioxide equivalents per kilogram of carcass during the final feeding stages, whereas beef raised with regenerative grazing resulted in a net *reduction* of 6 kilograms of carbon dioxide equivalents per kilogram of carcass.[26] These studies lead to the startling possibility that cattle can be raised under conditions that foster more carbon fixation than release and no soil erosion. At first glance this seems impossible, based on fundamental principles of carbon in the food chain, but skeptics have been taking another look. Scientists speculate that the regenerative grazing's conditions better approximate the natural behavior of a herd—feeding in dense groups for protection from predators and migrating frequently across the landscape—the behavior that produced the fertile Alfisols and Mollisols of Africa and North America before people perturbed the system. The increase in soil carbon is likely due to the plants remaining in the rapid-growth stage for longer than in conventional grazing. The plants are trimmed by the animals and then allowed to regrow, and the cycle is repeated, resulting in a net increase in photosynthesis compared with traditional grazing, where plants are nibbled down to nubs and require weeks of recovery to restart growth. More research is needed to determine the extent of the carbon benefits, but powerful evidence shows that these grazing practices are far better for the soil and carbon balance than traditional ones.

Critics of regenerative grazing point out that when the soil reaches its capacity for carbon storage, the process will no longer be a net carbon sink. But most soils that have been subjected to conventional animal husbandry have a long way to go before all of their carbon is restored to prior levels. And perhaps some soils can absorb even more carbon than they contained before people came along—like the Amazon Dark Earths and soils produced by plaggen agriculture on Papa Stour. In other words, we might carpet the Earth with great Anthrosols, soils that will be deeper and richer in carbon than the ones we inherited.

The Pink Houses in the East New York neighborhood of Brooklyn provide a vivid reminder that today's soil stewardship is not the exclusive domain of rural communities. The Pink Houses project is known for its bleak history of fifty years of neglect and poor management by the New York City Housing Authority followed by disinvestment, abandoning residents with dozens of unaddressed safety violations. But amid the 12 hectares of the dismal urban thicket of red brick buildings lies an unexpected sight—expanses of lush vegetable plots housed in raised beds tended by energetic residents. Gardeners distribute loads of black compost, plant diverse vegetables, hoe the beds to manage weeds, and harvest their bounty. In 2018, Pink Houses Community Farm distributed over 1,000 kilograms of food to over nine hundred residents, and in return the residents provided kitchen waste to produce compost that nourishes the soil.[27]

Well-managed urban farms like Pink Houses build soil through plant diversity, compost application, and rehabilitation of contaminated soil. Pink Houses is one of two urban farms in the East New York Farms! program. The farms are directed by Iyeshima Harris, a community activist who grew up in the Caribbean, where she connected with the land through her grandmother's garden. When she immigrated to the United States at the age of twelve, Iyeshima found herself boxed in by New York City's hard surfaces, which disconnected her from the earth, so she built a community garden at the back of her high school. While in college and working full time, Iyeshima scaled up her farming to East New York Farms! where she directs over forty staff and youth. Infected with Iyeshima's vision and energy, her staff and volunteers manage the gardens and sell their produce at the local farmers' market, which has celebrated food from the many cultures in the neighborhood for the past twenty years. Iyeshima Harris, full of passion and ideas, has become an icon of the East New York neighborhood, bringing farming and fresh produce to her community, nurturing soil, plants, and human spirits. To connect urban residents with soil and land, Iyeshima has to confront the strong association of farming with slavery that is common among Black city youth. She helps them replace the abhorrent images of the past with a new meaning—farming as a form of empowerment that enables people to reclaim their land and control their own food supply and nutrition.

The healthy soil of their urban farms in the East New York neighborhood is all the more surprising because it is in New York City, a place infamous for its toxic soil, contaminated with lead and other heavy metals. Lying quietly beneath Brooklyn's poisonous surface is some of the most valuable soil in the United States, which started to form about twenty thousand years ago when the last glacier left 100-meter-deep deposits of sand, silt, and rock that gave rise to today's soil. For decades, New York City construction projects have carted off 2–3 million tonnes of topsoil to toxic waste dump sites. In an innovative move, the city started the NYC Clean Soil Bank—the world's first municipal soil exchange—which supplies soil to local projects, including community gardens. The city removes the toxic layer and excavates the clean earth living underneath. In five years, the soil bank has delivered 500,000 tonnes of soil (enough to fill Yankee Stadium to a depth of 30 meters).[28] Among the beneficiaries are the Pink Houses community farmers, who also started their own clean soil bank to ensure their independence from the municipal program. Soil that was once buried, made unusable by a toxic barrier, has been brought to the surface, enriched with compost and good farming practices, and made fit to feed residents while building the Earth's carbon stockpile.

The urban gardens in New York are part of a worldwide movement to raise food and improve soil in cities, providing urbanites with locally produced food, a connection to the land, and a healthier environment. The incongruity of farming in concrete has captured the public's imagination. Today urban agriculture engages eight hundred million people worldwide who raise crops, honey, herbs, fruit, fish, and other animals for food. These innovative farmers build their enterprises in yards, at abandoned lots, on roofs, and in greenhouses to feed local people, often responding to the demands for nutritious produce in food deserts. They also employ local youth, teaching them agriculture skills and providing them green space. The farms range from microgardens in Jakarta to a 1.3-hectare garden on the roof of a warehouse in Paris. During the coronavirus pandemic, city farming boomed across Asia, feeding families from Singapore to India and inspiring experts to predict that by 2050 local production will top 180 million tonnes of food annually and feed two-thirds of the world's urban population. The future of urban farming is bright, improving human nutrition and urban life, bolstering soil, and capturing greenhouse gases.[29]

It is intriguing to imagine a new future for agriculture, one in which food is produced on rural land and in cities in backyards and on rooftops, stewarded by indigenous people with ancient practices and by those new to agriculture with conservation practices. Livestock would be raised with intensive rotational grazing to reduce their carbon footprint and soil depletion. In this version of the future, agriculture would *grow soil* in addition to crops, increasing carbon stocks, assuring future food production, and mitigating greenhouse gas emission. Farmers would be compensated for this service, so they wouldn't have to choose between their own survival and the Earth's. Farmers are already my heroes, but if they solved the soil and climate crises, they would be everyone's heroes.

10

A World With Soil

Cooperation is a most remarkable animal behavior. Bison form dense herds to deter predators; dolphins group around mothers during childbirth; pelicans congregate to concentrate fish and make hunting easier; honey bees shiver collectively to warm up their queen. And longhorn crazy ants collaborate to transport pieces of food that are far too large for one ant alone. People often use cooperation to make social change, but sometimes they don't, and the results are dismal. As we advance a new movement to protect soil, broad cooperation will be essential. Let's call this movement Save Our Soil (or SOS) and take a look at what's needed to turn an idea into a movement, first at the international level and then in the United States, as a case study for national soil policy.

The Save Our Soil movement needs to launch quickly to ensure that we have enough soil to sustain food production throughout the twenty-first century, but fortunately, the tools are available to create meaningful change. Soil-protective farming approaches used for thousands of years have been studied for the past century, so strategies for successful soil management are predicated on long experience and clear, compelling science—an ingredient lacking in many other environmental movements. Widespread use of no-till planting, cover crops, intercropping, water management, and intensive rotational grazing would bring most farmland erosion to a full stop. So the question is not how to save the soil but, rather, how to inspire or require people to take the actions that will save it. Starting at the international level, there is a great model of global cooperation—the Paris COP 21 climate talks. The cohesive voice in development of the international agreement on carbon emissions provided hope that the world might unite around a common concern, although the Paris Agreement was followed by the disappointment of limited follow-up and insufficient action. COVID-19 offers us another model of cooperation that produced tangible outcomes.

The actions of governments and behaviors of individuals during the COVID-19 pandemic make it clear that fast, cooperative action is possible. Despite the inadequate initial response of the U.S. government and several others to contain the virus, the cooperation was impressive. Scientists stepped up. The world community of coronavirus experts shared data and brainstormed solutions together, holding three-hour weekly symposia to discuss every aspect of the disease—the infection process, treatments, disinfectants, masks, and vaccines. Governments collaborated on strategies to contain the pandemic, and several private companies catapulted vaccines from development to distribution in mere months—an unprecedented feat supported by regulatory agencies, health care workers, and citizens. Although compliance with mandates to wear masks was maddeningly poor in some countries, the responsible people who faithfully wore them and adhered to social distancing guidelines saved millions of lives. Soil, too, deserves the type of cooperation that led to control of COVID-19.

Soil is a resource that is managed locally but is often claimed at a national level. Countries assume unquestioned dominion over their soil, considering it equivalent to land and territory. And yet soil flows freely in rivers and dust clouds, as we have seen with the transcontinental movement of Sahara sand and the Nile's transport of sediment across national borders, making its provenance hard to discern. In fact, no country owns its soil. And soil provides the platform for producing agricultural products—food, feed, and fiber—that are global commodities, frequently used on the opposite side of the world from where they were produced. The connection of soil with climate cements soil's position as a shared resource that should be managed in part through international cooperation.

Despite 120 years of brokering multilateral treaties about the environment, the international community has signed few agreements that focus on soil. The 1982 UN Environmental Programme World Soils Policy and the Food and Agriculture Organization World Soil Charter, also of 1982, along with the 1994 Convention on Desertification, recognized soil as a nonrenewable resource and delegated its preservation to each country with no international oversight. The treaties proposed no mechanisms for accountability. A 1991 European agreement among the countries transected by the Alps to save the mountain range's soil was adopted in 1998 and put into force in 2006. It recognized the fragile nature of the alpine landscape and proposed a soil conservation plan to ensure its sustainable use.[1] Although practices implemented after the treaty was signed have protected expanses of alpine soil, there is little evidence that steps would not have been taken without the treaty. The agreement to preserve the Alps provides a good model, however, that demonstrates how countries can collaborate to protect a regionally shared resource. Today we should move beyond regional agreements and advance a global declaration recognizing that soil is at once essential and imperiled and provides a tool to mitigate climate change. That declaration must also contain goals and measures for accountability.

Research shows that many people are overwhelmed with anxiety or even grief about climate change. Although each of us may be able to reduce our carbon footprint by conserving energy, individuals have few opportunities to *remove* greenhouse gases, which frustrates those who want to improve the climate outlook. These are the world's citizens who will be inspired by the chance to take affirmative action rather than simply being prohibited from certain activities. In a 2009 report prepared for the World Bank on the failure of the global response to climate change, Kari Marie Norgaard, a sociologist who studies attitudes toward climate change, described a collective "frustration and paralysis" that led to widespread climate apathy. She argued that people seek to preserve a sense of personal agency and control of their destiny and that studying the climate change problem pushes them the other way, toward feelings of helplessness, fear, and guilt. People also want to protect their sense of pride in their individual and national identities, which is diminished by admitting their own and their country's complicity in the climate disaster. Norgaard's work reveals two types of actions that people tend to avoid—contemplating risks to humans and engaging in behaviors that deviate from cultural norms, two aspects of social change that are worth considering in designing a plan.[2] Her work provides key signposts for developing policy. It is essential that international alliances and national strategies focus on positive action that can rehabilitate the earth beneath our feet, generate pride in our communities—whether local, national, or global—for implementing the strategies, and create new cultural norms. In turn, leaders at all levels need to focus on long-term strategies based on scientific evidence rather than quick fixes that may create a false sense of security that we are addressing the problem.

International treaties are a tool for conserving shared environmental resources. Creating a new compact for soil might be a reinvigorated version of 4p1000, the agreement introduced at the Paris climate talks in 2015. The limited support for 4p1000, indicated by only twenty-nine countries signing it, stemmed in part from its aspirational nature. Those who proposed 4p1000 did not really expect to increase the carbon content of every hectare on Earth by 0.4 percent per year, but others perceived it as a literal and therefore unreasonable goal. Some countries objected to the lack of specific targets or milestones. Although the text stated otherwise, the agreement's very title implied that all countries would be held to the same quantitative standard, leading to the perception that uniform goals would be imposed rather than inviting nations to identify goals and strategies that match their soil and its uses.[3]

So what would constitute a set of feasible goals for soil carbon sequestration? First, realistic goals must narrow the scope to agricultural land. Two post-Paris studies estimated that worldwide soil management changes have the potential to sequester an additional 2–3 gigatonnes of carbon per year in the top meter of agricultural soil, which would represent up to one-third of current emissions. Another study estimated that soil conservation practices on crop and grazing lands in the United States alone could annually sequester an additional 0.075

gigatonnes of carbon by 2025; a fourth argued that this was an over-estimate because it did not account for increased emissions that accompany expansion of certain agricultural endeavors, such as grazing animals.[4]

Eventually soil would reach its carbon-carrying capacity, but how much could be sequestered before that point is debated. Some researchers estimate that globally, soil can store an additional 0.4 gigatonnes per year at most. Critics of 4p1000 argue that soil has a finite carbon carrying capacity, and therefore once an equilibrium is achieved, the strategy will no longer work. If only that were our problem! If all agricultural soils worldwide reached their carbon-holding capacity, erosion's impact would be diminished to negligible, and greenhouse gases would be reduced. And there may be more potential for carbon sequestration than current estimates allow. Most estimates of the total carbon-carrying capacity do not consider the possibility that agricultural soil might be coaxed to hold more carbon than it did before it was cultivated. If it did, this would mean exceeding the soil health gap—the difference between carbon content on cultivated land and neighboring unmanaged land—and closing the gap has been suggested as the target for increased soil carbon.[5] The Amazon Dark Earths and plaggen agriculture in northern Europe are among examples of managed land that has a higher carbon content than surrounding untouched areas. The soil health gap provides a benchmark, but some regions might well exceed it.

Increasing carbon sequestration on the world's agricultural land by 0.4–1.0 gigatonne per year (using a conservative estimate) would mitigate 5–10 percent of current fossil fuel emissions of 9 gigatonnes per year. If the more optimistic estimate that we could sequester 3 gigatonnes per year is realistic, we might reach 30 percent of current emissions. Whether 5 percent or 30 percent of emissions, increased carbon sequestration must be only one of a slate of policies that re-

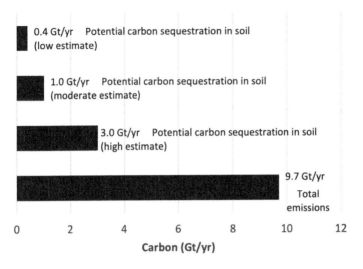

Figure 17. Potential carbon sequestration in soil compared with fossil fuel emissions. Illustration by Jo Handelsman.

duce emissions sufficiently to ensure that the atmosphere remains well under a devastating 2°C increase as stated in the Paris Agreement.[6] The opportunity for carbon sequestration in soil must not foster complacency about other climate mitigation research and activism (fig. 17).

Although projections about the rate and extent of carbon sequestration may vary, there is little disagreement about the fundamental premise that after losing 25 percent of its carbon since 1850, the Earth's soil is a carbon repository with untapped capacity.[7] Improving soil health is a worthy goal for the sake of soil. The ancillary benefit of mitigating a portion of greenhouse gas emissions—whether 5 percent or 30 percent—is compelling. Either goal is worth pursuing. Together they should be irresistible.

A revised 4p1000 should enable countries to adjust goals to match their soil types and use. Degraded land incorporates carbon more easily than healthy soil, and the carbon-carrying capacity of soils differs, so sequestration rates will differ among countries. The goal of 4p1000 should be reformulated, or replaced, by a global body convened in a manner similar to the COP 21 meeting. The new or revised treaty could focus on agricultural and agroforestry land, which is the source of extensive erosion and where we have the most control over carbon incorporation. The new agreement should propose feasible rather than aspirational goals, identify milestones, and provide flexibility for each country to set its own targets. With these modifications, more countries are likely to commit to it. It should be presented on the world stage with all the optimism it can incite, restoring to many of the world's citizens the hope, pride, and belief in restoring the health of the Earth that has been lost in dispiriting climate discussions.

Regardless of any international steps taken, it is vital that the United States undertake a strategic plan of action because of the nation's extensive agricultural footprint, high-quality and vulnerable soil, and high energy use. Although no single plan can fit all countries, the steps proposed here may serve as a case study to be adapted to other cultures and terrains.

Between 2005 and 2014, the organic carbon on managed U.S. cropland and grassland increased fivefold as a result of the Agriculture Department's Natural Resources Conservation Service programs that are designed to provide farmers with best practices to enhance soil health. Soil and climate expert Keith Paustian and his collaborators propose that the United States should accelerate carbon sequestration in soil to achieve another fivefold increase by 2025 and then accelerate it even more between 2025 and 2050. They argue that we need to take advantage of carbon sequestration potential in the top 20 centimeters of grassland and cropland soils, which could sequester up to 0.075 gigatonnes of carbon by 2025. Increasing soil carbon by 0.4 percent annually thereafter would incorporate 0.075 gigatonnes per year, equal to half of the nation's agricultural carbon footprint and the footprint of fourteen million U.S. residents.[8] This rate of carbon sequestration could continue until equilibrium has been reached. By then U.S. soil would be healthy, less erodible, and less demanding of fertilizers.

These goals need to be embedded in a national policy document. Surprisingly, the United States is one of only a few countries that lacks a strategic plan for soil. To develop a comprehensive plan to restore soil health and meet the goals of 4p1000 (should we arrive at an international agreement and ratify a new version), the federal government should build on the 2016 report *The State and Future of U.S. Soils: Framework for a Federal Strategic Plan for Soil Science,* which my office in OSTP requested from the Soil Science Interagency Working Group. The working group consisted of scientists from seven federal agencies as well as OSTP staffers Rich Pouyat and Parker Liautaud. By the end of President Obama's term, they had built a framework for a strategic plan. The framework included increasing public awareness about and appreciation of soil, delivering a set of best practices, setting targets for soil health, and improving methods to measure progress toward goals. The plan should focus on ways to build on the increases in soil carbon achieved with existing Natural Resources Conservation Service policies and aim to accelerate progress.[9]

The U.S. president should convene a coalition of nongovernmental actors to advise the Soil Science Interagency Working Group as it develops a detailed strategic plan for soil, as intended by the group that developed the framework. The complexity of soil and its role at the nexus of the food system, the environment, and agribusiness make it imperative to gather the wisdom and creativity of people who have diverse relationships with soil so that the plan addresses each group's constraints, needs, and potential contributions. Membership could be drawn from nonprofit organizations, consumers, farmers, indigenous peoples, and private industry.

While the government strategic planning process is under way, the broad coalition could design incentives for farmers to improve their practices and for retailers and consumers to support them. The coalition should consider launching a new label to identify food produced under soil-friendly conditions. A label such as "Produced by Carbon Heroes" would serve to link farmers, soil, and fighting climate change in everyone's minds. In just a few words it would recognize the effort and resources that farmers invest in the world's food supply. Just as organic-certified food carries a premium price tag, so could soil-safe food to compensate farmers for the initial costs of changing practices. Restaurants and food services in schools and other public institutions could demonstrate their commitment to protecting soil and mitigating climate change by serving only soil-safe food.

The stakeholder coalition should broker agreements from all sectors of the food system—particularly retailers and processors—to ensure that the costs of the label are shouldered by industrial partners as well as consumers. Partitioned responsibilities should be codified in legislation as a potential centerpiece for the upcoming Farm Bill, the legislation passed every five years that funds most agriculture- and food-related programs. The next bill, due in 2024, could reconfigure food stamp benefits to encourage purchase of soil-safe food without added cost to the recipients as was done in the Supplemental Nutrition Assistance Program to encourage purchase of produce. The Farm Bill or an executive order could require federal facilities to purchase soil-safe food for their cafeterias and dining halls.

Standards for soil-safe practices and a process to certify farms will be essential to success. Certification could be based on the practices used, such as some combination of cover crops, no-till planting, composting, strip cropping or intercropping, rotations, and regenerative grazing. Farmers might receive points for each practice implemented, or certification could be tied to demonstrating that soil carbon either increases over time or has reached equilibrium. Ensuring that the premium paid by consumers, retailers, and other members of the food system benefits farmers is essential for this strategy to be effective. By engaging a wide community in designing the incentives and cost-sharing for the label's benefits, the soil-safe label can be fortified by several sectors, making it robust and more financially beneficial to farmers than are current organic food labels.

Carbon markets are taking hold as a means to monetize prudent carbon management. Farmers could enter the market by earning credits based on the carbon content of their soil or the soil-safe practices they employ. The Agriculture Department could hold a carbon bank that would pay farmers for carbon sequestered in soil and sell the credits to corporations that need to offset their greenhouse gas emissions. A coalition of farm, food, and environmental groups called the Food and Agriculture Climate Alliance has endorsed this concept, and several corporations have begun to commercialize it. The startup company Indigo Ag, for example, is contracting with farmers to sequester carbon at fifteen dollars per ton. The international agricultural company Bayer started its Carbon Initiative, which will pay farmers for carbon sequestration practices.[10] Integration of carbon markets into agriculture is new but advancing rapidly. Carbon credits for farmers are likely to constitute the next national or international program to pay for soil health.

The meat industry is one of the most controversial aspects of the food supply. It raises ethical issues about treatment of animals, prompts environmental concerns about climate and pollution, and has troubling impacts on human health. But we can assume that meat production will continue for quite a while as domestic and international demand for it increases. Production of the corn and soybeans used as feed makes the meat industry destructive to midwestern soil. Similarly, traditional grazing systems make soils on the Great Plains more vulnerable to wind erosion. Both the environment and the industry could benefit from the label "Produced by Carbon Heroes." Meat producers have been excoriated by environmentalists because of cows' enormous carbon footprint, which results from their inefficient conversion of plant-based foods to meat and release of methane produced by the microorganisms in their rumens. Meat from cattle raised under an intensive rotational grazing regime that approaches carbon neutrality would be eligible for the label.[11] Grain fed to livestock would also need to meet soil-safe standards for the meat to be certified. The soil-safe label would benefit extensive midwestern farmland where corn and soybeans are grown for livestock feed as well as western grazing land. Reducing the damage done by continuous corn or even corn-soybean rotations would save tonnes of soil, particularly in the sloped land of Iowa, where soil is especially vulnerable to displacement by water. This change would not address other objections to meat production (animal welfare and human health concerns), but for those who choose to produce and consume meat, its damage to the soil and climate could be drastically reduced.

Crop insurance protects farmers from financial ruin in years of drought, flooding, or other events that cause catastrophic crop loss. Under the current model, the amount farmers are insured is based on the previous year's production, adjusted for changes in land area planted. To compensate farmers for switching to soil-conservation methods, the collaborative system of government and banks that currently provides crop insurance could reduce premiums for farms that steadily increase soil carbon or offer credit for soil-safe practices. The goal would be to make crop insurance a minimal cost to farmers practicing soil-safe farming. Moreover, crop insurance should be restructured to eliminate the current perverse disincentive for planting prairie strips. This is created by the use of the land area planted with the crop as the basis for coverage, which penalizes farmers for replacing 10 percent of their corn with prairie plantings. The strips and other soil-safe practices would diminish erosion and increase carbon, making the land less vulnerable to flood and drought damage, thereby reducing the need for payouts. Costs to insurers would decline, compensating them for receiving lower premiums from farmers.

Eventually, changes in crop insurance or food labeling could be embedded in the Farm Bill, but until then, the Department of Agriculture could enact a trial period using its prerogative to make policy within the broad strokes of the legislation that funds it. The Agriculture Department is a powerful and pivotal partner in launching a vigorous effort to save U.S. soil, but it shouldn't have to do it alone. The Natural Resource Conservation Service within USDA is dedicated to improving soil health and has already made great progress in increasing soil carbon through land conservation programs and education for farmers. Food labeling, crop insurance, and partnerships would provide additional levers to influence farm practice.

A food label and changes to crop insurance would compensate farmers for practicing soil conservation, but implementing these changes will require engagement across the food system and vigorous public support to motivate change. It's safe to say that most of us don't worry about soil and its fate every day, so it's important to raise public concern sufficiently to pressure lawmakers to legislate, government leaders to mold policy, food retailers to market soil-safe food, and consumers to buy it. This will require social change.

Harvard law professor Cass Sunstein has argued that all social change requires adjusting social norms. Once the norms are adjusted, the change spreads on its own as people pass along new beliefs, knowledge, habits, and action plans to others. Such change is advanced when we hear messages from many sources to reach a "tipping point," the idea introduced by Malcolm Gladwell to explain how some social changes suddenly catch on.[12] To achieve that point, societies need to remove barriers, provide incentives, and match messaging to desired outcomes. Discarding plastic was replaced by the recycling movement. Institutionalizing and marginalizing people with disabilities was replaced by accommodations and mainstreaming. Each of these changes was the result of several forces, including repeated exposure to data about the negative impact of the status quo; a strong voice advocating change; influential people joining the cause; laws and inducements; and the ease of adopting the new norm. Rarely does any one factor act alone: data without a charismatic voice falls flat; a single voice without reinforcement stops being heard; compelling arguments without an easy way for individuals to take action leave people frustrated.

The success of the organic food movement illustrates the impact of a tipping point. The movement started in the 1920s but didn't take hold in the United States until after publication of Rachel Carson's *Silent Spring,* which brought to light the environmental consequences of agricultural pesticides, herbicides, and fertilizers that were polluting waterways and entering the food supply. But Carson's book, as powerful as it was in generating a new lexicon and ideas about environmental protection among the public, did not spur a new movement alone. The effects of the insecticide DDT on the charismatic bald eagle became widely known. Then the Endangered Species Act was passed in 1967, Senator Gaylord Nelson introduced the first Earth Day in 1970 to elevate and increase public awareness about environmental issues, and three months later the Environmental Protection Agency was established by executive order with a mandate to protect human health and the environment from pollutants. In the 1970s, a counterculture movement advocated organic methods as part of environmentally sound lifestyles, and veterans returning from Vietnam joined the movement, protesting the use of herbicides after seeing the devastation caused by Agent Orange, a potent plant toxin, during the war.[13] Organic food remains an important but small part of U.S. agriculture. The annual market in the United States hit $20 billion during 2020, reaching a tipping point when it was supported by farmers, the food industry, and consumers. More sweeping and rapid change will be required to reach a tipping point for soil in time to save it.

Save Our Soil would need to engage farmers who produce grains, vegetables, and meat on all types of farms, large and small. The movement could promote the "Produced by Carbon Heroes" label and build public and consumer support. To maximize effectiveness, this effort should be based on research about making social change and avoiding pitfalls of previous movements.

Actions to restore soil should extend beyond farmers. A campaign to encourage installation of carbon-rich gardens in yards or on rooftops could engage millions of us with soil and the food system, thereby increasing appreciation for the natural resource and those who tend it. Community gardening programs, such as the East New York Farms!, could aid in forming cooperative groups who raise food and enrich soil.

The success of Save Our Soil will depend on a multipronged campaign that moves us to action by different appeals. A broad campaign would drive political activism, purchasing choices, and belief systems. Spreading awareness about the soil crisis and how to halt it will require information disseminated through the press, books, public service messages, social media, and school curricula. Which methods of communication have been used successfully in the past?

A proven avenue to inform people and influence behavior that is often ignored by people launching scientific initiatives is the arts and entertainment industry. A large body of social psychology literature shows that artistic messages that access emotional responses are effective behavior changers. Even as early as 1947, there was consensus that movies change behavior and that television can deliver public messages for society's benefit. For instance, following an episode of *Happy Days* (a mainstream television show in the 1970s) in which a popular character went to the library to meet girls, there was a 500 percent increase in applications for library cards among young people in the United States. Since then, the concept of entertainment education has been used to influence choices about substance abuse, childhood vaccination, and environmental protection. The strategy works. In combating drunk driving, for example, the concept of a designated driver was made commonplace in the United States within three years of inserting it into several television series.[14]

Various types of entertainment media should play an important role, as they have in the effort to build public concern about climate change. Al Gore's documentary *An Inconvenient Truth* provided a riveting, scientifically accurate portrait of the catastrophe that will result from mounting carbon emissions. The 2004 fictional blockbuster film *The Day After Tomorrow* reached ten times as many people as Gore's documentary with a bold portrait of climate change. Using stunning special effects and sometimes preposterous scientific premises, the film was both lauded and condemned by scientists. It reached more people than any other climate message and became a reference point for climate change discussions. Even articles in respected scientific journals referred to the film, and some scientists appreciated it for extending the conversation to a broader audience and increasing willingness to take action to avert global warming. Other scientists could not forgive the scientific inaccuracies, worrying that the film would plant misconceptions in the minds of the public or even increase skepticism about climate change. Still others argued that the film served a purpose as a rhetorical device to stimulate discussion. And some believed that the film's gender and race stereotypes counteracted its benefits by promulgating a troubling worldview.[15] Perhaps soil activists can improve on the climate change movement's use of entertainment media by commissioning a scientifically accurate and spellbinding box office hit movie that will introduce soil and its preservation as a commonplace topic of conversation.

Video games are another popular medium that can use emotional engagement to spread the word about soil. In 2020, about 75 percent of people in the United States played video games, whereas only half of U.S. adults had read a nonfiction book in the last year.[16] With three billion gamers worldwide, video games are a powerful tool. Who would have thought that a game about plants entitled *Plants vs. Zombies* could attract eight million players and rank thirty-third among most popular video games? Its wild success led one of my colleagues, psychology professor Karen Schloss, to propose a video game about soil erosion. In addition to films and video games, the Save Our Soil movement should use social media, YouTube videos, and public service announcements to present the public with scientifically responsible, factual accounts and fictional narratives about soil. Diverse types of media will reach a broad cross-section of society across age, employment, socioeconomic class, and lifestyle.

The stakeholder coalition could lead the information campaign, connecting with diverse industries and messengers. The nature of messaging is critical. The post-COVID-19 world is fatigued with crises. People want to connect with others and be part of making change that will have a noticeable impact. An international movement needs to engage broadly, making us feel part of something larger than ourselves, giving us a sense of unified purpose and connection to the world. The messages to invite everyone's participation and ownership of Save Our Soil should focus on concrete individual action that can address the soil problem as well as the challenge that people have found so overwhelming and paralyzing—climate change—by supporting efforts to increase soil carbon. Providing personal agency will energize each of us and bring hope. Messaging should balance the negative consequences of inaction at a global level with positive outcomes of collective action. Given the choice of carbon- and soil-friendly food, many of us will be ready to pay a premium to feel that we are contributing to solving the soil and climate crises and supporting farmers working toward those goals. The movement's messaging should emphasize action, agency, and pride in being part of a national or global force for change.

The Save Our Soil movement must be careful not to overpromise (adding carbon to soil is not a panacea for either erosion or climate change), so that its goals are achieved. Intermediate successes could be celebrated each year on December 5, International Soil Day. The president should include updates on U.S. progress toward soil carbon goals in the State of the Union address so that the public anticipates hearing about soil from the president just as they do other urgent issues of the day.

Many countries have launched wartime campaigns that engaged almost every citizen. During World War II, for instance, American women worked in munitions factories and planted Victory Gardens, and British and Australian women joined the Women's Land Armies to fill the farm labor gap. We need that type of broad-based engagement in a common goal to conquer the soil crisis.

People across the world may appreciate the freshness of the message about the shared fix for soil erosion and climate change. Soil cannot fix global warming by itself, but soil management can contribute to a multifaceted climate agenda. The newness of the approach of sequestering carbon in soil, instead of focusing attention exclusively on ways to reduce fossil fuel emissions, may empower and thrill a public that is eager to play a role in fixing a crisis rather than being at the mercy of others who may or may not consider the climate a priority. Introducing them to a new crisis, soil erosion, along with a strategy to address it immediately may also spread optimism and appreciation for human beings' ability to solve problems.

The ancient nature of the solution should reassure us that the knowledge is in hand to replenish the soil without causing other environmental disasters. The understandable, simple nature of the fix will engage those who might be repelled by high-tech fixes such as geoengineering or manipulating nutrients in the sea and other extreme proposals for addressing climate change that have potential for dangerous unintended consequences.[17] Incentives, broad messaging campaigns, and legislation can make the fix a reality.

When I worked for President Obama, his staff printed little cards that carried inspiring quotes from the president and distributed them to the White House staff. I keep one above my writing desk that reads, "Everything we do needs to be infused with a sense of possibility. We are not scared of the future." That captured the hope that the president instilled in all of us. It might sound like empty optimism, but he made sure we all remembered how much we had accomplished against tough odds. He spoke often about the great strides our country has made and the power of the human spirit and ingenuity. I hope that after reading these pages, you share that optimism. Although nothing is a panacea and no single approach will stop soil erosion and climate change, we have figured out how to mitigate these two great threats to our species and the Earth. Although the data should scare us, we should not be scared of the future. Human ingenuity combined with the regenerative power of soil can save us from a precarious future.

Acknowledgments

Many people helped me during the forty years I spent preparing to write this book. The soil scientists, climate change experts, policy wonks, writers, and so many others shared, debated, taught, and helped me become the author of this book.

This would be a different book without the research and collaboration of Kayla Cohen, whose keen insights, exhaustive research, and meticulous editing shaped the content and style. Her skill with crafting language and lyrical ear are evident on every page. My gratitude is beyond words for the gifts she brought to this book. And thanks to Pam and David Cohen for all the support and encouragement they provided Kayla and to Vicki Chandler for introducing us.

I also wish to express particular gratitude to:

Parker Liautaud and Andrew Hanus, who worked tirelessly on soil policy with me in the White House for their fabulous research and helping to shape my ideas about soil; and I thank Parker for suggesting the title;

Elizabeth Stulberg, for her work on soil in the White House and beyond and her insights into carbon policy;

Rick Cruse, for his contributions to my White House soil policy development, his willingness to pick up the phone, visit, or write; Rick transformed my thinking about modern soil science; during the writing of this book, Rick was an endless source of information and ideas, responding to late night emails even from his remote cabin in the wilds of Iowa; I'll always be grateful that he inspired me to write this book and for his feedback as it was written;

Garth Harmsworth, Jessica Hutchings, William Wetere, and Tui Aroha War-
menhoven, who provided essential insight into Māori culture, history,
and soil management; Garth also provided a great many photographs
and reviewed and improved the Māori section;

Anabel Ford, for her assistance with research on the Maya and review of the
Maya section and for challenging prevailing dogma;

John Valley, Iyeshima Harris, Georgianna Scott, and Joe Bragger, for open-
ing their lives to Kayla and me, sharing their experience and expertise,
reviewing sections of the manuscript, and allowing us to feature them;

Alfred Hartemink and Bradley Miller, for always being willing to fill gaps in
my soil knowledge and for reviewing sections of the manuscript;

Anton Petrus, for rushing to the Chernozem fields outside of Kiev to snap
photos of the soil before it was obscured by the corn crop;

Keith Paustian, for help with the intricacies of carbon balance;

Gina Caison, for insights about soil in fiction;

Matthew Ruark, Michael Bell, William Tracy, Corbett Grainger, Erin Silva,
Anita Chan, Jeremy Teperman, James Farquhar, James Kasting, Alison
Gale, Kevin Masarik, Amin Emadi, Matt Seib, Michael Parsen, Jim
Hebbe, Jeanne Whitish, William Gartner, William Wetere, Karin Rem-
melzwaal, David Browning, Andrew W. Stevens, and Carrie Laboski,
for helping with many aspects of research on soil and policy;

Dara Park, Rama, Dwight Sipler, Matthias Vanmaercke, Paul Reich, Alfred
Hartemink, the National Soil Survey Center, the Monticello/Thomas
Jefferson Foundation, Landcare Research NZ Ltd, USDA-NRCS, and
GRID-Arendal, for providing photographs and maps;

Liz Edwards, Sophie Wolfson, Bill Nelson, Helen Jones, Bobbi Angell, and
Marc G. Chevrette, for their wonderful and creative illustrations, and
to Liz for her singular chapter page drawings;

Bob Prior at MIT Press, for his encouragement to write this book;

Martel DenHartog, for her encouragement, attention to detail, and bib-
liographic assistance, down to the wire;

Laura Langley, for protecting my writing time and support of every type;

The Handelsman Lab, for always providing me with data fixes and being my
friends;

ACKNOWLEDGMENTS

Elizabeth Sylvia and staff at Yale University Press, for advice and creative contributions;

Jean Thomson Black, editor extraordinaire at Yale University Press, for sharing her brilliance, deep experience, and breadth of knowledge about science and publishing; for her staunch support and for shepherding the book through every joyful and painful step with grace and humor;

Hilary and Alix Handelsman, my sisters, who edited tirelessly and loved me throughout;

and Casey, thank you for everything.

Abbreviations

FAO	Food and Agriculture Organization of the United Nations
IPCC	Intergovernmental Panel on Climate Change
ITPS	Intergovernmental Technical Panel on Soils
PNAS	*Proceedings of the National Academy of Sciences of the United States*
UNESCO	United Nations Educational, Scientific and Cultural Organization
USDA	United States Department of Agriculture
USGS	United States Geological Survey

Notes

Prologue

1. FAO, *Healthy Soils Are the Basis for Healthy Food Production* (Rome: FAO, 2015).

2. David A. N. Ussiri and Rattan Lal, *Carbon Sequestration for Climate Change Mitigation and Adaptation* (Cham, Switzerland: Springer International, 2017), 80, 86.

3. FAO and ITPS, *Status of the World's Soil Resources: Main Report* (Rome: FAO, 2015), 101–103; Ronald Amundson et al., "Soil and Human Security in the 21st Century," *Science* 348 (2015): 1261071; David R. Montgomery, "Soil Erosion and Agricultural Sustainability," *PNAS* 104 (2014): 13268–13272; Stanley W. Trimble, *Man-Induced Soil Erosion of the Southern Piedmont, 1700–1970* (Ankeny, Iowa: Soil and Water Conservation Society, 2008); Richard Cruse et al., "Daily Estimates of Rainfall, Water Runoff, and Soil Erosion in Iowa," *Journal of Soil and Water Conservation* 61 (2006): 191, pl. 6; Evan A. Thaler, Isaac J. Larsen, and Qian Yu, "The Extent of Soil Loss Across the US Corn Belt," *PNAS* 118 (2021): e1922375118.

4. Montgomery, "Soil Erosion."

5. "Welcome to the '4 per 1000' Initiative," 4 per 1000, https://www.4p1000.org.

Chapter 1. Dawning—An Invisible Crisis

1. USDA, *Summary Report: 2012 National Resources Inventory* (Washington, D.C.: Natural Resources Conservation Service; and Ames, Iowa: Center for Survey Statistics and Methodology, 2015); Jesse Newman, Renée Rigdon, and Patrick McGroarty, "The World's Appetite Is Threatening the Mississippi River," *Wall Street Journal,* July 2, 2019, http://graphics.wsj.com/mississippi/.

Chapter 2. Earth's Dark Matter

1. J. W. Valley, "A Cool Early Earth?," *Scientific American* 293 (2005): 58–63.

2. Tara Djokic et al., "Earliest Signs of Life on Land Preserved in ca. 3.5 Ga

Hot Spring Deposits," *Nature Communications* 8, no. 15263 (2017); Takayuki Tashiro et al., "Early Trace of Life from 3.95 Ga Sedimentary Rocks in Labrador, Canada," *Nature* 549 (2017): 516–518.

3. Eiichi Tajika and Mariko Harada, "Great Oxidation Event and Snowball Earth," in *Astrobiology: From the Origins of Life to the Search for Extraterrestrial Intelligence,* ed. Akihiko Yamagishi, Takeshi Kakegawa, and Tomohiro Usui (Singapore: Springer Nature Singapore, 2019), 261–271.

4. On zircon as "Earth's timekeeper," see Valley, "Cool Early Earth?," 64. On zircon dating, see Simon A. Wilde et al., "Evidence from Detrital Zircons for the Existence of Continental Crust and Oceans on the Earth 4.4 Gyr Ago," *Nature* 409 (2001): 175–178.

5. Wilde et al., "Evidence from Detrital Zircons."

6. J. William Schopf, *Cradle of Life: The Discovery of Earth's Earliest Fossils* (Princeton, N.J.: Princeton University Press, 1999), 5.

7. J. William Schopf et al., "SIMS Analyses of the Oldest Known Assemblage of Microfossils Document Their Taxon-Correlated Carbon Isotope Compositions," *PNAS* 115 (2018): 53.

8. Harvinder Singh, *Steel Fiber Reinforced Concrete: Behavior, Modelling and Design* (Singapore: Springer Singapore, 2017), 2.

9. Larry Horath, *Fundamentals of Materials Science for Technologists: Properties, Testing, and Laboratory Exercises,* 3rd ed. (Long Grove, Ill.: Waveland, 2019), 165; Bradley D. Fahlman, "Solid-State Chemistry," in *Materials Chemistry,* 3rd ed. (Dordrecht: Springer Netherlands, 2018).

10. Hans-Curt Flemming and Stefan Wuertz, "Bacteria and Archaea on Earth and Their Abundance in Biofilms," *Nature Reviews Microbiology* 17 (2019): 247–260.

11. Yinon M. Bar-On, Rob Phillips, and Ron Milo, "The Biomass Distribution on Earth," *PNAS* 115 (2018): 6506; Laureano A. Gherardi and Osvaldo E. Sala, "Global Patterns and Climatic Controls of Belowground Net Carbon Fixation," *PNAS* 117 (2020): 20038–20043; Birgit W. Hütsch, Jürgen Augustin, and Wolfgang Merbach, "Plant Rhizodeposition: An Important Source for Carbon Turnover in Soils," *Journal of Plant Nutrition and Soil Science* 165 (2002): 397–407; Christophe Nguyen, "Rhizodeposition of Organic C by Plants: Mechanisms and Controls," *Agronomy* 23 (2003): 375–396; Hans Lambers, "Growth, Respiration, Exudation and Symbiotic Associations: The Fate of Carbon Translocated to the Roots," in *Root Development and Function,* ed. P. J. Gregory, J. V. Lake, and D. A. Rose (Cambridge:

Cambridge University Press, 1987), 125–145; Rajeew Kumar, Sharad Pandey, and Apury Pandey, "Plant Roots and Carbon Sequestration," *Current Science* 91 (2006): 885–890.

12. "Mount St. Helens: From the 1980 Eruption to 2000," U.S. Geological Survey Fact Sheet 036-00, USGS, last modified March 1, 2005, https://pubs.usgs.gov/fs /2000/fs036-00/.

13. A. H. Fitter et al., "Biodiversity and Ecosystem Function in Soil," *Functional Ecology* 19 (2005): 369–377; Thibaud Decaëns, "Macroecological Patterns in Soil Communities," *Global Ecology and Biogeology* 19 (2010): 287–302; Richard D. Bargdett and Wim H. van der Putten, "Belowground Biodiversity and Ecosystem Functioning," *Nature* 515 (2014): 505–511; Alan Kergunteuil et al., "The Abundance, Diversity, and Metabolic Footprint of Soil Nematodes Is Highest in High Elevation Alpine Grasslands," *Frontiers in Ecology and Evolution* 4 (2016): 84; Tom Bongers and Marina Bongers, "Functional Diversity of Nematodes," *Applied Soil Ecology* 10 (1998): 239–251; Patrick D. Schloss and Jo Handelsman, "Toward a Census of Bacteria in Soil," *PLoS Computational Biology* 2 (2006): e92; Alberto Orgiazzi et al., *Global Soil Biodiversity Atlas* (Luxembourg: Publications Office of the European Union, 2015); Noah Fierer, "Earthworms' Place on Earth," *Science* 366 (2019): 425–426.

Chapter 3. Earth Works

1. Christian Feller, Lydie Chapuis-Lardy, and Fiorenzo Ugolini, "The Representation of Soil in the Western Art: From Genesis to Pedogenesis," in *Soil and Culture,* ed. Edward R. Landa and Christian Feller (Dordrecht: Springer Netherlands, 2009), 3–22; "Bhudevi," New World Encyclopedia, https://www.newworld encyclopedia.org/entry/Bhudevi; Hassan El-Ramady et al., "Soils and Human Creation in the Holy Quran from the Point of View of Soil Science," *Environmental Biodiversity and Soil Security* 3 (2019): 2–3.

2. Ernest Thompson Seton and Julia M. Seton, comps., *The Gospel of the Redman,* commemorative ed. (Bloomington, Ind.: World Wisdom, 2005), 80.

3. Martin K. Jones and Xinyi Liu, "Origins of Agriculture in East Asia," *Science* 324 (2009): 730–731; Ainit Snir et al., "The Origin of Cultivation and Proto-Weeds, Long Before Neolithic Farming," *PLoS ONE* 10 (2015): e0131422.

4. Jeanne Sept, "Early Hominin Ecology," in *Basics in Human Evolution,* ed. Michael P. Muehlenbein (Amsterdam: Elsevier, 2015), 86–101; Ewen Callaway, "Oldest *Homo sapiens* Fossil Claim Rewrites Our Species' History," *Nature News,* June 8, 2017, https://www.nature.com/news/oldest-homo-sapiens-fossil-claim-rewrites-our

-species-history-1.22114; Brigitte M. Holt, "Anatomically Modern *Homo sapiens,*" in *Basics in Human Evolution,* ed. Michael P. Muehlenbein (Amsterdam: Elsevier, 2015), 177; Nicholas Toth and Kathy Schick, "Overview of Paleolithic Archaeology," in *Handbook of Paleoanthropology,* ed. Winfried Henke and Ian Tattersall (Berlin: Springer, 2015), 2441–2464; Ansley J. Coale, "The History of the Human Population," *Scientific American* 231 (1974): 40–51.

5. Sanjai J. Parikh and Bruce R. James, "Soil: The Foundation of Agriculture," *Nature Education Knowledge* 3 (2012): 2; Mark B. Tauger, "The Origins of Agriculture and the Dual Subordination," in *Agriculture in World History* (London: Routledge, 2010), 3–14; Jeffrey P. Severinghaus and Edward J. Brook, "Abrupt Climate Change at the End of the Last Glacial Period Inferred from Trapped Air in Polar Ice," *Science* 286 (1999): 930; Snir et al., "Origin of Cultivation."

6. FAO, *Healthy Soils Are the Basis for Healthy Food Production* (Rome: FAO, 2015).

7. R. L. Holle and R. E. López, "A Comparison of Current Lightning Death Rates in the U.S. with Other Locations and Times" (paper presented at International Conference on Lightning and Static Electricity, Royal Aeronautical Society, Blackpool, England, 2003), paper 103-34.

8. Edwin B. Fred, Ira L. Baldwin, and Elizabeth McCoy, *Root Nodule Bacteria and Leguminous Plants,* University of Wisconsin Studies in Science, no. 5 (1932), 4.

9. W. M. Stewart et al., "The Contribution of Commercial Fertilizer Nutrients to Food Production," *Agronomy* 97 (2005): 1.

10. Birgit W. Hütsch, Jürgen Augustin, and Wolfgang Merbach, "Plant Rhizodeposition: An Important Source for Carbon Turnover in Soils," *Journal of Plant Nutrition and Soil Science* 165 (2002): 397–407.

11. David J. Levy-Booth et al., "Cycling of Extracellular DNA in the Soil Environment," *Soil Biology and Biochemistry* 39 (2007): 2977–2991; G. Pietramellara et al., "Extracellular DNA in Soil and Sediment: Fate and Ecological Relevance," *Biology and Fertility of Soils* 45 (2009): 219–235; Ohana Y. A. Costa, Jos M. Raaijmakers, and Eiko E. Kuramae, "Microbial Extracellular Polymeric Substances: Ecological Function and Impact on Soil Aggregation," *Frontiers in Microbiology* 9 (2018): 1636.

12. On plant and fungi interactions, see B. Wang and Y.-L. Qiu, "Phylogenetic Distribution and Evolution of Mycorrhizas in Land Plants," *Mycorrhiza* 16 (2006): 300, 353. On phosphorus fertilizer reduction, see David R. Montgomery, *Dirt: The Erosion of Civilizations, with a New Preface* (Berkeley: University of California Press, 2012), 187–188.

13. Judith D. Schwartz, "Soil as Carbon Storehouse: New Weapon in Climate Fight?," *Yale Environment 360,* March 4, 2014; Rattan Lal, "Soil Carbon Sequestration to Mitigate Climate Change," *Geoderma* 123 (2004): 1–22.

14. On groundwater supply and use, see *The United Nations World Water Development Report 2015: Water for a Sustainable World: Facts and Figures* (Paris: UNESCO, 2015), 2, 9, http://www.unesco.org/new/fileadmin/MULTIMEDIA/HQ/SC/images /WWDR2015Facts_Figures_ENG_web.pdf; "World's Groundwater Resources Are Suffering from Poor Governance, Experts Say," Media Services, UNESCO, May 5, 2012, http://www.unesco.org/new/en/media-services/single-view/news/worlds_ground water_resources_are_suffering_from_poor_gove. On groundwater use in India, see Himanshu Kulkarni, Mihir Shah, and P. S. Vijay Shankar, "Shaping the Contours of Groundwater Governance in India," *Journal of Hydrology: Regional Studies* 4, part A (2015): 173. On well water in the United States and Canada, see "Water Sources," Drinking Water, CDC, last reviewed April 10, 2009, https://www.cdc.gov/healthy water/drinking/public/water_sources.html#one; and Expert Panel on Groundwater, *The Sustainable Management of Groundwater in Canada* (Ottawa, Ont.: Council of Canadian Academies, 2009), 3, http://www.cec.org/wp-content/uploads/wpallimport /files/17-1-sub-appendix_ix_-_expert_panel_on_groundwater_-_sustainable_man agement_of_groundwater_-_2009.pdf.

15. National Research Council, *The New Science of Metagenomics: Revealing the Secrets of Our Microbial Planet* (Washington, D.C.: National Academies Press, 2007), 19.

16. *World Water Development Report 2015,* 6.

17. Vigdis Torsvik and Lise Øvreås, "Microbial Diversity and Function in Soil: From Genes to Ecosystems," *Current Opinion in Microbiology* 5 (2002): 240; Vigdis Torsvik, Jostein Goksøyr, and Frida Lise Daae, "High Diversity in DNA of Soil Bacteria," *Applied and Environmental Microbiology* 56 (1990): 782.

18. Hannah Ritchie, "How Do We Reduce Antibiotic Resistance from Livestock?," Our World in Data, November 16, 2017, https://ourworldindata.org/antibiotic -resistance-from-livestock.

19. Amanda Hurley et al., "Tiny Earth: A Big Idea for STEM Education and Antibiotic Discovery," *mBio* 12 (2021): e03432-20.

20. Richard H. Baltz, "Marcel Faber Roundtable: Is Our Antibiotic Pipeline Unproductive Because of Starvation, Constipation or Lack of Inspiration?," *Journal of Industrial and Microbial Biotechnology* 33 (2006): 507–513; Uddhav K. Shigdel et al., "Genomic Discovery of an Evolutionarily Programmed Modality for Small-

Molecule Targeting of an Intractable Protein Surface," *PNAS* 117 (2020): 17195–17203.

21. Pasquale Borrelli et al., "An Assessment of the Global Impact of 21st Century Land Use Change on Soil Erosion," *Nature Communications* 8 (2017): 2013.

Chapter 4. Chaos to Orders

1. Hans Jenny, *Factors of Soil Formation: A System of Quantitative Pedology* (New York: McGraw-Hill, 1941), 12.

2. Pavel Krasilnikov et al., eds., *A Handbook of Soil Terminology, Correlation and Classification* (London: Routledge, 2009), 1–2.

3. John King, "Plants Are Cool, but Why?," in *Reaching for the Sun: How Plants Work* (Cambridge: Cambridge University Press, 1997), 3; Robert S. Wallace, "Record-Holding Plants," *Plant Sciences,* Encyclopedia.com, https://www.encyclopedia.com/science/news-wires-white-papers-and-books/record-holding-plants.

4. Soil Survey Staff, *Illustrated Guide to Soil Taxonomy,* ver. 2 (Lincoln, Nebr.: U.S. Department of Agriculture, Natural Resources Conservation Service, National Soil Survey Center, 2015), 2–5.

5. Krasilnikov et al., *Handbook,* 2; Martin K. Jones and Xinyi Liu, "Origins of Agriculture in East Asia," *Science* 324 (2009): 730–731; Ainit Snir et al., "The Origin of Cultivation and Proto-Weeds, Long Before Neolithic Farming," *PLoS ONE* 10 (2015): e0131422; David C. Coleman, D. A. Crossley Jr., and Paul F. Hendrix, "1—Historical Overview of Soils and the Fitness of the Soil Environment," in *Fundamentals of Soil Ecology,* 2nd ed. (Amsterdam: Elsevier Academic Press, 2004), 2.

6. Edmund Ruffin, *An Essay on Calcareous Manures* (Petersburg, Va.: J. W. Campbell, 1832); Stanley W. Buol et al., *Soil Genesis and Classification,* 4th ed. (Ames: Iowa State University Press, 1999), 9; Vasily A. Esakov, "Dokuchaev, Vasily Vasilievich," *Complete Dictionary of Scientific Biography,* Encyclopedia.com, https://www.encyclopedia.com/people/science-and-technology/environmental-studies-biographies/vasily-vasilievich-dokuchaev.

7. On Soviet era soil classification, see Krasilnikov et al., *Handbook,* 11. On French soil classification, see Commision de Pédologie et de Cartographie des Sols, *Classification des sols* (1967), https://horizon.documentation.ird.fr/exl-doc/pleins_textes/divers16-03/12186.pdf; Freddy O. Nachtergaele, "New Developments in Soil Classification: The World Reference Base for Soil Resources," in *Quatorzième Réunion du Sous-Comité ouest et centre africain de corrélation des sols pour la mise en valeur des terres* (Rome: FAO, 2002), 25; and "Soil," BGR, https://www.bgr.bund.de/EN/Themen

/Boden/boden_node_en.html. On the World Reference Base, see Jozef Deckers et al., *World Reference Base for Soil Resources—in a Nutshell,* European Soil Bureau, Research Report no. 7 (2001), 173.

8. L. T. West, M. J. Singer, and A. E. Hartemink, eds., "Introduction," in *The Soils of the USA* (Cham, Switzerland: Springer, 2017), 2–3, fig. 1.1.

9. On the twelve orders of soil, see Pan Min Huang, Yuncong Li, and Malcolm E. Sumner, eds., *Handbook of Soil Sciences Properties and Processes,* 2nd ed. (Boca Raton, Fla.: CRC Press, 2012); Stanley W. Buol et al., *Soil Genesis and Classification,* 6th ed. (West Sussex, UK: Wiley-Blackwell, 2011); Stanley W. Buol et al., *Soil Genesis and Classification,* 5th ed. (Ames: Iowa State University Press, 2003); "The Twelve Soil Orders," Global Rangelands, https://globalrangelands.org/topics/rangeland-ecology/twelve-soil-orders#Inceptisols; and "Inceptisols," University of Idaho, https://www.uidaho.edu/cals/soil-orders/inceptisols.

Chapter 5. Wind, Water, and Plows

1. Rattan Lal and William C. Moldenhauer, "Effects of Soil Erosion on Crop Productivity," *Critical Reviews in Plant Sciences* 5 (1987): 303–367.

2. Rattan Lal, "Soil Erosion and Gaseous Emissions," *Applied Sciences* 10 (2020): 1–13; G. A. Fox et al., "Reservoir Sedimentation and Upstream Sediment Sources: Perspectives and Future Research Needs on Streambank and Gully Erosion," *Environmental Management* 57 (2016): 945–955; "Hypoxia in the Gulf of Mexico," U.S. Department of the Interior, USGS, last modified October 23, 2017, https://toxics.usgs.gov/hypoxia/mississippi/oct_jun/index.html.

3. FAO and ITPS, *Status of the World's Soil Resources: Main Report* (Rome: FAO, 2015), 103, 177; Dan Pennock, *Soil Erosion: The Greatest Challenge to Sustainable Soil Management* (Rome: FAO, 2019), 3; Christoffel den Biggelaar et al., "Crop Yield Losses to Soil Erosion at Regional and Global Scales: Evidence from Plot-Level and GIS Data," in *Land Quality, Agricultural Productivity, and Food Security: Biophysical Processes and Economic Choices at Local, Regional, and Global Levels,* ed. Keith Wiebe (Cheltenham, UK: Edward Elgar, 2003), 271; David R. Montgomery, "Soil Erosion and Agricultural Sustainability," *PNAS* 104 (2014): 13268–13272.

4. T. E. Fenton, M. Kazemi, and M. A. Lauterbach-Barrett, "Erosional Impact on Organic Matter Content and Productivity of Selected Iowa Soils," *Soil and Tillage Research* 81 (2005): 163–171; Lal, "Soil Erosion."

5. On arid soils and dust emissions, see Sujith Ravi et al., "Aeolian Processes and the Biosphere," *Reviews of Geophysics* 49 (2011): 1; Paul Reich, Hari Eswaran, and Fred

Beinroth, "Global Dimensions of Vulnerability to Wind and Water Erosion," in *Sustaining the Global Farm: Selected Papers from the 10th International Soil Conservation Organization Meeting, May 24–29, 1999*, ed. D. E. Stott, R. H. Mohtar, and G. C. Steinhardt (West Lafayette, Ind.: International Soil Conservation Organization in cooperation with the USDA and Purdue University, 2001), 838–846; Frank E. Urban et al., "Unseen Dust Emission and Global Dust Abundance: Documenting Dust Emission from the Mojave Desert (USA) by Daily Remote Camera Imagery and Wind-Erosion Measurements," *Journal of Geophysical Research: Atmospheres* 123 (2018): 8735–8753; Yaping Shao et al., "Dust Cycle: An Emerging Core Theme in Earth System Science," *Aeolian Research* 2 (2011): 182; Paul Ginoux et al., "Global-Scale Attribution of Anthropogenic and Natural Dust Sources and Their Emission Rates Based on MODIS Deep Blue Aerosol Products," *Review of Geophysics* 50 (2012): RG3005; and FAO and ITPS, *Status,* 101. On wind erosion in the United States, see USDA, *Summary Report: 2012 National Resources Inventory* (Washington, D.C.: Natural Resources Conservation Service; and Ames, Iowa: Center for Survey Statistics and Methodology, 2015), 2–8.

6. Ryan Schleeter, "The Grapes of Wrath," *National Geographic,* April 7, 2014, https://www.nationalgeographic.org/article/grapes-wrath/; Timothy Egan, *The Worst Hard Time: The Untold Story of Those Who Survived the Great American Dust Bowl* (New York: Houghton Mifflin Harcourt, 2006), 198–221.

7. Dong Zhibao, Wang Xunming, and Liu Lianyou, "Wind Erosion in Arid and Semiarid China: An Overview," *Journal of Soil and Water Conservation* 55 (2000): 439–444; "The Expansion of the Gobi Desert," ESRI, https://www.arcgis.com/apps/MapJournal/index.html?appid=c108d6ff4937464f86cb0fbef796f515; FAO and ITPS, *Status,* 290; Xunming Wang et al., "Desertification in China: An Assessment," *Earth Science Reviews* 88 (2008): 188–206; Wang Tao, "Aeolian Desertification and Its Control in Northern China," *International Soil and Water Conservation Research* 2 (2014): 35.

8. Sarah Gibbens, "Why This Dust Storm in India Turned Deadly," *National Geographic,* May 3, 2018, https://www.nationalgeographic.com/news/2018/05/india-dust-storm-wind-fatalities-science-spd/; India Today Web Desk, "Thunderstorm Hits Delhi-NCR: How Man's Neglect for Soil Management Has Given Rise to a Monster," India Today, May 3, 2018, https://www.indiatoday.in/education-today/gk-current-affairs/story/dust-storm-death-toll-facts-on-dust-storm-html-1225662-2018-05-03; Tapan J. Purakayastha et al., "Soil Resources Affecting Food Security and Safety in South Asia," in *World Soil Resources and Food Security,* ed. Rattan Lal and B. A. Stewart (Boca Raton, Fla.: CRC Press, 2012), 276.

9. FAO and ITPS, *Status,* 101.

10. J. D. Walsh et al., "Our Changing Climate," in *Climate Change Impacts in the United States: The Third National Climate Assessment,* ed. Jerry M. Melillo, Terese Richmond, and Gary W. Yohe (Washington, D.C.: U.S. Global Change Research Program, 2014), 19–67.

11. Thomas Schumacher et al., "Modeling Spatial Variation in Productivity Due to Tillage and Water Erosion," *Soil and Tillage Research* 51 (1999): 331–339; Pennock, *Soil Erosion,* 2.

12. On water erosion in sub-Saharan Africa, see FAO and ITPS, *Status,* 247–248. On salination in India, see G. Swarajyalakshmi, P. Gurumurthy, and G. V. Subbaiah, "Soil Salinity in South India: Problems and Solutions," *Journal of Crop Production* 7 (2003): 247–275.

13. Alexsey Sidorchuk and Valentin Nikolaevich Golosov, "Erosion and Sedimentation on the Russian Plain, II: The History of Erosion and Sedimentation During the Period of Intensive Agriculture," *Hydrological Processes* 17 (2003): 3347–3358; John M. Laflen and Dennis C. Flanagan, "The Development of U.S. Soil Erosion Prediction and Modeling," *International Soil and Water Conservation Research* 1 (2013): 1–11, 2.

14. Jessica J. Veenstra and C. Lee Burras, "Soil Profile Transformation After 50 Years of Agricultural Land Use," *Soil Science Society of America Journal* 79 (2015): 1154–1162.

15. Y. P. Hsieh, K. T. Grant, and G. C. Bugna, "A Field Method for Soil Erosion Measurements in Agricultural and Natural Lands," *Journal of Soil and Water Conservation* 64 (2009): 374; Lal, "Soil Erosion"; A. Mahmoudzadeh, Wayne D. Erskine, and C. Myers, "Sediment Yields and Soil Loss Rates from Native Forest, Pasture, and Cultivated Land in the Bathurst Area, New South Wales," *Australian Forestry* 65 (2002): 73–80.

16. "Ending Nuclear Testing," International Day Against Nuclear Tests, United Nations: 29 August, United Nations, https://www.un.org/en/observances/end-nuclear-tests-day/history; V. A. Kashparov et al., "Soil Contamination with 90Sr in the Near Zone of the Chernobyl Accident," *Journal of Environmental Radioactivity* 56 (2001): 285–298; Paolo Porto et al., "Validating Erosion Rate Estimates Provided by Caesium-137 Measurements for Two Small Forested Catchments in Calabria, Southern Italy," *Land Degradation and Development* 14 (2007): 389–408; Eric W. Portenga and Paul R. Bierman, "Understanding Earth's Eroding Surface with [10]Be," *Geological Society of America Today* 21 (2011): 4–10.

17. C. King et al., "The Application of Remote-Sensing Data to Monitoring

and Modelling of Soil Erosion," *Catena* 62 (2005): 79–93; Anton Vrieling, "Satellite Remote Sensing for Water Erosion Assessment: A Review," *Catena* 65 (2006): 2–18; Mehrez Zribi, Nicolas Baghdadi, and Michel Nolin, "Remote Sensing of Soil," *Applied and Environmental Soil Science* (2011): 1–2; "Landsat 8," Landsat Missions, USGS, https://www.usgs.gov/core-science-systems/nli/landsat/landsat-8?qt-science _support_page_related_con=0#qt-science_support_page_related_con; Marián Jenčo et al., "Mapping Soil Degradation on Arable Land with Aerial Photography and Erosion Models, Case Study from Danube Lowland, Slovakia," *Remote Sensing* 12 (2020): 1–17.

18. A. W. Zingg, "Degree and Length of Land Slope as It Affects Soil Loss in Runoff," *Agricultural Engineering* 21 (1940): 59–64; Walter H. Wischmeier, "A Rainfall Erosion Index for a Universal Soil-Loss Equation," *Soil Science Society America* 23 (1959): 246–249; Nyle C. Brady, *The Nature and Properties of Soil*, 8th ed. (New York: Macmillan, 1974), 639; Malcolm Newson, *Land, Water and Development: Sustainable Management of River Basin Systems*, 2nd ed. (London: Routledge, 1997), 218; Walter H. Wischmeier and Dwight D. Smith, *Predicting Rainfall-Erosion Losses from Cropland East of the Rocky Mountains: Guide for Selection of Practices for Soil and Water Conservation* (Washington, D.C.: Agricultural Research Service, USDA, in cooperation with Purdue Agricultural Experiment Station, 1965), 47; Laflen and Flanagan, "Development," 1–11; National Research Council, *Soil Conservation: Assessing the National Resources Inventory*, vol. 1 (Washington, D.C.: National Academies Press, 1986), 59; Christine Alewell et al., "Using the USLE: Chances, Challenges, and Limitations of Soil Erosion Modelling," *International Soil and Water Conservation Research* 7 (2019): 203–225; Fox et al., "Reservoir Sedimentation," 945–955; J. Poesen, D. Torri, and T. Vanwalleghem, "Chapter 19—Gully Erosion: Procedures to Adopt When Modelling Soil Erosion in Landscapes Affected by Gullying," in *Handbook of Erosion Modelling*, ed. R. P. C. Morgan and M. A. Nearing (Oxford: Blackwell-Wiley, 2011); National Research Council, *Soil Conservation: An Assessment of the National Resources Inventory*, vol. 2 (Washington, D.C.: National Academies Press, 1986).

19. Dennis C. Flanagan, "Modeling Soil and Water Conservation," in *Soil and Water Conservation: A Celebration of 75 Years*, ed. Jorge A. Delgado, Clark J. Gantzer, and Gretchen F. Sassenrath (Ankeny, Iowa: Soil and Water Conservation Society, 2020), 255–269; Brian Gelder et al., "The Daily Erosion Project: Daily Estimates of Water Runoff, Soil Detachment, and Erosion," *Earth Surface Processes and Landforms* 43 (2018): 1105–1117.

20. Stanley W. Trimble and Pierre Crosson, "US Soil Erosion Rates: Myth and Reality," *Science* 289 (2000): 248–250; Laflen and Flanagan, "Development," 1–11.

21. A GIS (geographic information system) is a framework for gathering, managing, and analyzing data; see "What Is GIS?," esri, https://www.esri.com/en-us/what-is-gis/overview. Aafaf El Jazouli et al., "Soil Erosion Modeled with USLE, GIS, and Remote Sensing: A Case Study of Ikkour Watershed in Middle Atlas (Morocco)," *Geoscience Letters* 4, no. 25 (2017); D. P. Shrestha, M. Suriyaprasit, and S. Prachansri, "Assessing Soil Erosion in Inaccessible Mountainous Areas in the Tropics: The Use of Land Cover and Topographic Parameters in a Case Study in Thailand," *Catena* 121 (2014): 40–52; Sohan Kumar Ghimire, Daisuke Higaki, and Tara Prasad Bhattarai, "Estimation of Soil Erosion Rates and Eroded Sediment in a Degraded Catchment of the Siwalik Hills, Nepal," *Land* 2 (2013): 370–391.

22. On Fiji, see FAO and ITPS, *Status,* 487. On the United States and specifically Iowa, see USDA and Iowa State University, *2015 National Resources Inventory: Summary Report* (Washington, D.C.: Natural Resources Conservation Service and Center for Survey Statistics and Methodology, 2018), 5–37; Craig Cox, Andrew Hug, and Nils Bruzelius, *Losing Ground* (Washington, D.C.: Environmental Working Group, April 2011), 13; and Bradley Miller, "Physiography of Iowa," Geospatial Laboratory for Soil Informatics, Iowa State University, December 23, 2020. Evan A. Thaler, Isaac J. Larsen, and Qian Yu, "The Extent of Soil Loss Across the US Corn Belt," *PNAS* 118 (2021): e1922375118.

23. Thomas Jefferson to Charles W. Peale, 1813, in *Thomas Jefferson's Garden Book,* ed. E. M. Betts (Monticello, Va.: Thomas Jefferson Foundation, 1999), 509.

24. David B. Grigg, *The Agricultural Systems of the World: An Evolutionary Approach* (London: Cambridge University Press, 1974), 256–283.

25. R. A. Houghton, "The Annual Net Flux of Carbon to the Atmosphere from Changes in Land Use, 1850–1990," *Tellus B: Chemical and Physical Meteorology* 51 (1999): 298–313; Eric A. Davidson and Ilse L. Ackerman, "Changes in Soil Carbon Inventories Following Cultivation of Previously Untilled Soils," *Biogeochemistry* 20 (1993): 161–193.

26. "Rate of Deforestation," Global Challenges, The World Counts, https://www.theworldcounts.com/challenges/planet-earth/forests-and-deserts/rate-of-deforestation/story; David R. Montgomery, *Dirt: The Erosion of Civilizations,* 2nd ed. (Berkeley: University of California Press, 2012), 49–81.

27. Lucas Reusser, Paul Bierman, and Dylan Rood, "Quantifying Human Impacts on Rates of Erosion and Sediment Transport at a Landscape Scale," *Geology* 43 (2015): 171–174; R. B. Daniels, "Soil Erosion and Degradation in the Southern Pied-

mont of the USA," in *Land Transformation in Agriculture,* ed. M. G. Wolman and F. G. A. Fournier (New York: John Wiley and Sons, 1987), 407–428.

28. Steven Davies, "Estimated Population of American Colonies: 1610 to 1780," Vancouver Island University, https://web.viu.ca/davies/H320/population.colonies .htm; Nicolas A. Jelinski et al., "Meteoric Beryllium-10 as a Tracer of Erosion Due to Postsettlement Land Use in West-Central Minnesota, USA," *Journal of Geophysical Research: Earth Surface* 124 (2019): 874–901; Bruce H. Wilkinson and Brandon J. McElroy, "The Impact of Humans on Continental Erosion and Sedimentation," *Geological Society of America Bulletin* 119 (2007): 140–156; Reich, Eswaran, and Beinroth, "Global Dimensions," 838–846; Montgomery, "Soil Erosion," 13268–13272.

29. Xiaobing Liu et al., "Overview of Mollisols in the World: Distribution, Land Use and Management," *Canadian Journal of Soil Science* 92 (2011): 383–402.

30. R. Skuodienė and Donata Tomchuk, "Root Mass and Root to Shoot Ratio of Different Perennial Forage Plants Under Western Lithuania Climatic Conditions," *Romanian Agricultural Research* 32 (2015); Sergi Munne-Bosch, "Perennial Roots to Immortality," *Plant Physiology* 166 (2014): 720–725; M. A. Bolinder et al., "Root Biomass and Shoot to Root Ratios of Perennial Forage Crops in Eastern Canada," *Canadian Journal of Plant Science* 82 (2002): 731–737.

31. Xiaochao Chen et al., "Changes in Root Size and Distribution in Relation to Nitrogen Accumulation During Maize Breeding in China," *Plant Soil* 374 (2014): 121–130; J. Giles Waines and Bahman Ehdaie, "Domestication and Crop Physiology: Roots of Green-Revolution Wheat," *Annals of Botany* 100 (2007): 991–998; Meghann E. Jarchow and Matt Liebman, "Tradeoffs in Biomass and Nutrient Allocation in Prairies and Corn Managed for Bioenergy Production," *Crop Science* 52 (2012): 1330–1342; Qiuying Tian et al., "Genotypic Difference in Nitrogen Acquisition Ability in Maize Plants Is Related to the Coordination of Leaf and Root Growth," *Journal of Plant Nutrition* 29 (2006): 317–330; Rex D. Pieper, "Chapter 6—Grasslands of Central North America," in *Grasslands of the World,* ed. J. M. Suttie, S. G. Reynolds, and C. Batello (Rome: FAO, 2005), 221–263.

32. Thomas Jefferson to Tristan Dalton, 1817, in *Thomas Jefferson's Garden Book,* ed. E. M. Betts (Monticello, Va.: Thomas Jefferson Foundation, 1999), 570.

33. S. G. Whisenant, *Repairing Damaged Wildlands: A Process-Oriented, Landscape-Scale Approach* (New York: Cambridge University Press, 1999); S. G. Whisenant, "Terrestrial Systems," in *Handbook of Ecological Restoration,* vol. 1, ed. M. R. Perrow and A. J. Davy (New York: Cambridge University Press, 2002), 83–105; Elizabeth G. King and Richard J. Hobbs, "Identifying Linkages Among Conceptual Models of

Ecosystem Degradation and Restoration: Towards an Integrative Framework," *Restoration Ecology* 14 (2006): 369–378.

34. Eric F. Lambin and Patrick Meyfroidt, "Global Land Use Change, Economic Globalization, and the Looming Land Scarcity," *PNAS* 108 (2011): 3465–3472.

35. Shaochuang Liu et al., "Pinpointing the Sources and Measuring the Lengths of the Principal Rivers of the World," *International Journal of Digital Earth* 2 (2009): 80–87.

36. Maurice L. Schwartz, ed., *Encyclopedia of Coastal Science* (Dordrecht: Springer Netherlands, 2005), 358.

37. Waleed Hamza, "The Nile Delta," in *The Nile,* ed. H. J. Dumont (Dordrecht: Springer Netherlands, 2009), 75–94; Scott W. Nixon, "Replacing the Nile: Are Anthropogenic Nutrients Providing the Fertility Once Brought to the Mediterranean by a Great River?," *AMBIO: A Journal of the Human Environment* 32 (2003): 30–39.

38. James P. M. Syvitski et al., "Impact of Humans on the Flux of Terrestrial Sediment to the Global Coastal Ocean," *Science* 308 (2005): 376–380; Khalid Mahmood, *Reservoir Sedimentation: Impact, Extent, and Mitigation* (Washington, D.C.: International Bank for Reconstruction and Development, 1987); Schwartz, *Encyclopedia of Coastal Science,* 358; Committee on Cost Savings in Dams, "Cost Savings in Dams (Draft of ICOLD Bulletin)," HydroCoop, http://www.hydrocoop.org/dams-cost-savings-icold/.

39. Walsh et al., "Our Changing Climate."

40. Simon Michael Papalexiou and Alberto Montanari, "Global and Regional Increase of Precipitation Extremes Under Global Warming," *Water Resources Research* 55 (2019): 4901–4914; IPCC, *Climate Change and Land: An IPCC Special Report on Climate Change, Desertification, Land Degradation, Sustainable Land Management, Food Security, and Greenhouse Gas Fluxes in Terrestrial Ecosystems* (2019), 6–7, https://www.ipcc.ch/srccl/.

41. Papalexiou and Montanari, "Global and Regional Increase"; IPCC, *Climate Change,* 6–7, 45.

42. Jock R. Anderson and Jesuthason Thampapillai, *Soil Conservation in Developing Countries: Project and Policy Intervention* (Washington, D.C.: World Bank, 1990), 17; Jelinski et al., "Meteoric Beryllium-10," 874–901; Chris Arsenault, "Only 60 Years of Farming Left If Soil Degradation Continues," *Scientific American,* December 5, 2014, https://www.scientificamerican.com/article/only-60-years-of-farming-left-if-soil-degradation-continues/.

Chapter 6. Rocky Planet

1. International Organization for Migration and United Nations Convention to Combat Desertification, *Addressing the Land Degradation-Migration Nexus: The Role of the UNCCD* (Geneva: International Organization for Migration, 2019); Ephraim Nkonya et al., "Global Cost of Land Degradation," in *Economics of Land Degradation and Improvement: A Global Assessment for Sustainable Development,* ed. Ephraim Nkonya, Alisher Mirzabaev, and Joachim von Braun, 117–165 (Cham, Switzerland: Springer International, 2016), 156; "Media Release: Worsening Worldwide Land Degradation Now 'Critical,' Undermining Well-Being of 3.2 Billion People," IPBES, released March 23, 2018, https://ipbes.net/news/media-release-worsening -worldwide-land-degradation-now-%E2%80%98critical%E2%80%99-undermining -well-being-32.

2. FAO and ITPS, *Status of the World's Soil Resources: Main Report* (Rome: FAO, 2015), 176.

3. Evan A. Thaler, Isaac J. Larsen, and Qian Yu, "The Extent of Soil Loss Across the US Corn Belt," *PNAS* 118 (2021): e1922375118.

4. Jonathan A. Foley et al., "Solutions for a Cultivated Planet," *Nature* 478 (2011): 337–342; Katherine Tully et al., "The State of Soil Degradation in Sub-Saharan Africa: Baselines, Trajectories, and Solutions," *Sustainability* 7 (2015): 6523–6562; I. I. Obiadi et al., "Gully Erosion in Anambra State, South East Nigeria: Issues and Solutions," *International Journal of Environmental Sciences* 2 (2011): 802; Babatunde J. Fagbohun et al., "GIS-Based Estimation of Soil Erosion Rates and Identification of Critical Areas in Anambra Sub-Basin, Nigeria," *Modeling Earth Systems and Environment* 2, no. 159 (August 2016); Benedicta Dike et al., "Potential Soil Loss Rates in Urualla, Nigeria Using RUSLE," *Global Journal of Science Frontier Research* 18, no. 2 (2018).

5. J. S. C. Mbagwu, Rattan Lal, and T. W. Scott, "Effects of Desurfacing of Alfisols and Ultisols in Southern Nigeria: I. Crop Performance," *Soil Science Society of America Journal* 48 (1984): 828–833.

6. Jude Nwafor Eze, "Drought Occurrences and Its Implications on the Households in Yobe State, Nigeria," *Geoenvironmental Disasters* 5, no. 18 (October 2018); R. Osabohien, E. Osabuohien, and E. Urhie, "Food Security, Institutional Framework, and Technology: Examining the Nexus in Nigeria Using ARDL Approach," *Current Nutrition and Food Science* 4, no. 2 (2018): 154–163; Esther Ngumbi, "To Ensure Food Security, Keep Soils Healthy" (blog), *World Policy,* December 12, 2017, http://worldpolicy.org/2017/12/12/to-ensure-food-security-keep-soils-healthy/;

Food Security Information Network, *2019 Global Report on Food Crises: Joint Analysis for Better Decisions* (Washington, D.C.: International Food Policy Research Institute, 2019), 18.

7. FAO, *Conservation des sols et des eaux dans les zones semi-arides* (Rome: FAO, 1990), 6; Mohamed Yjjou et al., "Modélisation de L'érosion Hydrique via les SIG et L'équation Universelle des Pertes en Sol au Niveau du Bassin Versant de l'Oum Er-Rbia," *The International Journal of Engineering and Science* 3, no. 8 (2014): 83; "Morocco Economic Outlook," African Development Bank Group, accessed January 7, 2021, https://www.afdb.org/en/countries/north-africa/morocco/morocco -economic-outlook.

8. Oliver Kiptoo Kirui and Alisher Mirzabaev, "Economics of Land Degradation in Eastern Africa" (working paper, ZEF Working Paper Series No. 128, Center for Development Research (ZEF), University of Bonn, 2014), 1; Addis Ababa, "Growth and Transformation Plan (GTP) 2010/11–2014/15" (draft, Ministry of Finance and Economic Development, 2010); Mahmud Yesuf, Salvatore Di Falco, et al., "The Impact of Climate Change and Adaptation on Food Production in Low-Income Countries: Evidence from the Nile Basin, Ethiopia" (discussion paper, International Food Policy Research Institute, 2008); Paschal Assey et al., *Environment at the Heart of Tanzania's Development: Lessons from Tanzania's National Strategy for Growth and Reduction of Poverty (MKUKUTA)* (London: International Institute for Environment and Development, 2007); Ritu Verma, *Gender, Land, and Livelihoods in East Africa: Through Farmers' Eyes* (Ottawa: International Development Research Centre, 2001); Abhijit Banerjee and Esther Duflo, *Poor Economics* (New York: Public Affairs, 2011), 134–135, 138.

9. Martin Khor, "Land Degradation Causes $10 Billion Loss to South Asia Annually," *Global Policy Forum,* https://www.globalpolicy.org/global-taxes/49705-land -degradation-causes-10-billion-loss-to-south-asi; FAO and ITPS, *Status of the World's Soil Resources: Main Report* (Rome: FAO, 2015); Dipak Sarkar et al., eds. *Strategies for Arresting Land Degradation in South Asian Countries* (Dhaka: SAARC Agriculture Centre, 2011), 38, 48.

10. "Bhutan: Committed to Conservation," World Wildlife Foundation, https:// www.worldwildlife.org/projects/bhutan-committed-to-conservation.

11. "Improved Maize Varieties and Partnerships Welcomed in Bhutan," CIMMYT E-News, International Maize and Wheat Improvement Center, May 14, 2012, https:// www.cimmyt.org/news/improved-maize-varieties-and-partnerships-welcomed-in -bhutan/; Karma Dema Dorji, "Strategies for Arresting Land Degradation in Bhu-

tan," in *Strategies for Arresting Land Degradation in South Asian Countries,* ed. Dipak Sarkar et al. (Dhaka, Bangladesh: SAARC Agricultural Centre, 2011), 59–71; Karma Wangdi Y and Rudra Bahadur Shrestha, "Family Farmers' Cooperatives Towards Ending Poverty and Hunger in Bhutan," in *Family Farmers' Cooperatives: Ending Poverty and Hunger in South Asia,* ed. Rudra Bahadur Shrestha et al. (Bangladesh: SAARC Agriculture Center, Philippines: Asian Farmers' Association, and India: National Dairy Development Board, 2020), 49; Royal Government of Bhutan, *Bhutan: In Pursuit of Sustainable Development,* National Report for the United Nations Conference on Sustainable Development 2012, https://sustainabledevelopment.un.org /content/documents/798bhutanreport.pdf.

12. Royal Government of Bhutan, *Bhutan;* Ephraim Nkonya et al., "Economics of Land Degradation and Improvement in Bhutan," in *Economics of Land Degradation and Improvement—A Global Assessment for Sustainable Development,* ed. Ephraim Nkonya, Alisher Mirzabaev, and Joachim von Braun (Washington, D.C.: Springer International Publishing, 2016), 327–383; United Nations Development Program and Global Environment Facility, *Bhutan: National Action Program to Combat Land Degradation,* 2009, https://www.acauthorities.org/sites/aca/files/country doc/Bhutan%20National%20Action%20Program%20to%20Combat%20Land%20 Degradation.pdf; Sangay Wangchuk and Stephen F. Siebert, "Agricultural Change in Bumthang, Bhutan: Market Opportunities, Government Policies, and Climate Change," *Society and Natural Resources: An International Journal* 26 (2013): 1375–1389.

13. Robert Repetto, "Soil Loss and Population Pressure on Java," *AMBIO: A Journal of the Human Environment* 15 (1986): 14–18; Iwan Rudiarto and W. Doppler, "Impact of Land Use Change in Accelerating Soil Erosion in Indonesian Upland Area: A Case of Dieng Plateau, Central Java—Indonesia," *International Journal of AgriScience* 3, no. 7 (July 2013): 574; Anna Strutt, "Trade Liberalisation and Soil Degradation in Indonesia," in *Indonesia in a Reforming World Economy: Effects on Agriculture, Trade and the Environment,* ed. Kym Anderson et al. (South Australia: University of Adelaide Press, 2009): 40–60; Salahudin Muhidin, "Population Projections in Indonesia During the 20th Century," in *The Population of Indonesia* (Amsterdam: Rozenberg, 2002), 90; Bram Peper, "Population Growth in Java in the 19th Century," *Journal of Demography* 24, no. 1 (1970).

14. FAO, *Small Family Farms Country Factsheet: Indonesia* (Rome: FAO, 2018); Diane Perrons, *Globalization and Social Change: People and Places in a Divided World* (Routledge, 2004), 92.

15. Atieno Mboya Samandari, *Gender-Responsive Land Degradation Neutrality* (working paper, Land Outlook, United Nations Convention to Combat Desertification, 2017), 3–15, https://knowledge.unccd.int/sites/default/files/2018-06/3.%20 Gender-Responsive%2BLDN__A_M__Samandari.pdf.

16. FAO, *Smallholders and Family Farmers,* 2012, http://www.fao.org/fileadmin /templates/nr/sustainability_pathways/docs/Factsheet_SMALLHOLDERS.pdf.

17. Ivan Franko, "Chernozems of Ukraine: Past, Present, and Future Perspectives," *Soil Science Annual* 70 (2019): 193–197.

18. Timothy Snyder, *Black Earth: The Holocaust as History and Warning* (New York: Tim Duggan Books, 2016); Turi Fileccia et al., *Ukraine: Soil Fertility to Strengthen Climate Resilience* (Rome: FAO, 2014).

19. "Soil Fertility to Increase Climate Resilience in Ukraine," The World Bank, December 5, 2015, https://www.worldbank.org/en/news/feature/2014/12/05/ukraine -soil; "Ukraine, FAO Unite to Save Healthy Soil," FAO, May 24, 2019, http://www .fao.org/europe/news/detail-news/en/c/1195526/; "FAO Launches Training Courses to Help Farmers Stop Land Degradation in Ukraine," FAO, February 19, 2019, http://www.fao.org/europe/news/detail-news/en/c/1180938/.

20. Xiobang Liu et al., "Overview of Mollisols in the World: Distribution, Land Use and Management," *Canadian Journal of Soil Science* 92 (2011): 383–402; H. H. Bennett, "The Cost of Soil Erosion," *Ohio Journal of Science* 33 (1933): 271–279; David Pimentel et al., "Environmental and Economic Costs of Soil Erosion and Conservation Benefits," *Science* 267, no. 5201 (1995): 1120.

21. Tiago Santos Telles et al., "Valuation and Assessment of Soil Erosion Costs," *Scientia Agricola* 70, no. 3 (2013).

22. National Agricultural Statistics Service, "2017 Census of Agriculture," USDA, 1–6, https://www.nass.usda.gov/Publications/AgCensus/2017/Full_Report/Volume _1,_Chapter_1_State_Level/Iowa/iarefmap.pdf.

23. Donnelle Eller, "Erosion Estimated to Cost Iowa $1 Billion in Yield," *Des Moines Register,* May 3, 2014, https://www.desmoinesregister.com/story/money /agriculture/2014/05/03/erosion-estimated-cost-iowa-billion-yield/8682651/; "Ukraine, FAO Unite to Save Healthy Soil," FAO; Craig Cox, Andrew Hug, and Nils Bruzelius, *Losing Ground* (Washington, D.C.: Environmental Working Group, April 2011), 13; Dennis B. Egli and Jerry L. Hatfield, "Yield and Yield Gaps in Central U.S. Corn Production Systems," *Agronomy Journal* 106 (March 2014): 2248–2254; Richard M. Cruse, *Economic Impacts of Soil Erosion in Iowa* (Leopold Center Completed Grant Reports, 2016); *2019 Iowa Farm Costs and Returns,* Ag Decision Maker (Iowa State

University Extension and Outreach, 2020); Yanru Liang et al., "Impacts of Simulated Erosion and Soil Amendments on Greenhouse Gas Fluxes and Maize Yield in Miamian Soil of Central Ohio," *Scientific Reports* 8, 520 (January 2018).

24. Cox, Hug, and Bruzelius, *Losing Ground,* 13.

25. National Marine Fisheries Service, *Fisheries Economics of the United States, 2015,* May 2017, National Oceanic and Atmospheric Association, https://www.fish eries.noaa.gov/feature-story/fisheries-economics-united-states-2015; Mississippi River/ Gulf of Mexico Watershed Nutrient Task Force, "Implementing the HTF 2008 Action Plan," Environmental Protection Agency, https://www.epa.gov/ms-htf/imple menting-htf-2008-action-plan; Environmental Protection Agency, *Protecting and Preserving the Gulf of Mexico: 2017 Annual Report,* 2017; Mary Caperton Morton, "Gulf Dead Zone Looms Large in 2019," *EOS* 100 (July 2019); Sergey S. Rabotyagov et al., "Cost-Effective Targeting of Conservation Investments to Reduce the Northern Gulf of Mexico Hypoxic Zone," *PNAS* 111 (2014): 18530–18535.

26. London Gibson and Sarah Bowman, "Disappearing Beaches, Crumbling Roads: Lake Michigan Cities Face 'Heartbreaking' Erosion," *Indianapolis Star,* March 24, 2020, https://www.indystar.com/story/news/environment/2020/03/24/lake-michigan -cities-indiana-struggle-heartbreaking-erosion/5031489002/.

27. Orlando Milesi and Marianela Jarroud, "Soil Degradation Threatens Nutrition in Latin America," *Inter Press Service,* June 15, 2016, http://www.ipsnews.net /2016/06/soil-degradation-threatens-nutrition-in-latin-america/; Karl S. Zimmerer, "Soil Erosion and Labor Shortages in the Andes with Special Reference to Bolivia, 1953–91: Implications for 'Conservation-with-Development,'" *World Development* 21 (1993): 1659–1675; Annemieke de Kort, "Soil Erosion Assessment in the Dryland Areas of Bolivia Using the RUSLE 3D Model" (MA thesis, Wageningen University, 2013), https://edepot.wur.nl/278541.

28. Pasquale Borrelli et al., "An Assessment of the Global Impact of 21st Century Land Use Change on Soil Erosion," *Nature Communications* 8 (December 2017); André Almagro et al., "Projected Climate Change Impacts in Rainfall Erosivity over Brazil," *Scientific Reports* 7 (August 2017); PwC Brazil, *Agribusiness in Brazil: An Overview,* 2013, 3, https://www.pwc.com.br/pt/publicacoes/setores-atividade /assets/agribusiness/2013/pwc-agribusiness-brazil-overview-13.pdf; Viviana Zalles, "Near Doubling of Brazil's Intensive Row Crop Area Since 2000," *PNAS* 116 (2019): 428–435; Gustavo H. Merten and Jean P. G. Minella, "The Expansion of Brazilian Agriculture: Soil Erosion Scenarios," *International Soil and Water Conservation Research* 1 (2013): 37–48; Cristian Youlton et al., "Changes in Erosion and Runoff Due to Re-

placement of Pasture Land with Sugarcane Crops," *Sustainability* 8 (2016): 685; Nilo S. F. de Andrade et al., "Economic and Technical Impact in Soil and Nutrient Loss Through Erosion in the Cultivation of Sugar Cane," *Engenharia Agrícola* 31 (2011): 539–550; Tiago Santos Telles et al., "The Costs of Soil Erosion," *Revista Brasileira de Ciência do Solo* 35 (2011): 287–298.

29. David Pimentel and Michael Burgess, "Soil Threatens Food Production," *Agriculture* 3 (2013): 443–463; Chris Arsenault, "Only 60 Years of Farming Left If Soil Degradation Continues," *Scientific American,* December 5, 2014, https://www.scientificamerican.com/article/only-60-years-of-farming-left-if-soil-degradation-continues/; FAO, "International Year of Soil Conference," 2015 Year of Soils, July 6, 2015, http://www.fao.org/soils-2015/events/detail/en/c/338738/; UN General Assembly, "Food Production Must Double by 2050 to Meet Demand from World's Growing Population, Innovative Strategies Needed to Combat Hunger, Experts Tell Second Committee," UN Meetings Coverage and Press Releases, October 9, 2009, https://www.un.org/press/en/2009/gaef3242.doc.htm.

30. Prabhu L. Pingali, "Green Revolution: Impacts, Limits, and the Path Ahead," *PNAS* 109 (2012): 12302–12308; "Annual Yield of Rice in India from Financial Year 1991 to 2018, with an Estimate for 2019," Statista, https://www.statista.com/statistics/764299/india-yield-of-rice/; R. L. Nielsen, "Historical Corn Grain Yields in the U.S.," Purdue University, updated April 2020, https://www.agry.purdue.edu/ext/corn/news/timeless/YieldTrends.html.

31. Deepak K. Ray et al., "Recent Patterns of Crop Yield Growth and Stagnation," *Nature Communications* 3 (2012): 1293; Zvi Hochman, David L. Gobbert, and Heidi Horan, "Climate Trends Account for Stalled Wheat Yields in Australia Since 1990," *Global Change Biology* 23 (2017): 2071–2081; Bernhard Schauberger et al., "Yield Trends, Variability, and Stagnation Analysis of Major Crops in France over More Than a Century," *Scientific Reports* 8 (2018): 16865; Peter Crosskey, "UK 'Yield Plateau' for Wheat and Colza," Agricultural and Rural Convention 2020, January 15, 2013, https://www.arc2020.eu/uk-yield-plateau-for-wheat-and-colza/.

32. Christine Kinealy, "Saving the Irish Poor: Charity and the Great Famine," *The 1846–1851 Famine in Ireland: Echoes and Repercussions,* Cahiers du MIMMOC (December 2015), https://doi.org/10.4000/mimmoc.1845; Ed O'Loughlin and Mihir Zaveri, "Irish Return an Old Favor, Helping Native Americans Battling the Virus," *New York Times,* May 5, 2020.

33. Committee on Commodity Problems, *Historical Background on Food Aid and Key Milestones* (Rome: FAO, 2005); "International Food Aid After 50 Years: A

Brief History of Modern Food Aid," Cornell University, updated May 20, 2011, https://www.cornell.edu/video/international-food-aid-2-brief-history-of-modern -food-aid; Cynthia Graber and Nicola Twilley, "How the U.S. Became the World's Largest Food-Aid Donor," *Atlantic,* May 23, 2018, https://www.theatlantic.com /health/archive/2018/05/how-the-us-became-the-worlds-largest-food-aid-donor /560951/; "A Short History of U.S. International Food Assistance," U.S. Department of State, https://2009-2017.state.gov/p/eur/ci/it/milanexpo2015/c67068.htm; "Famine," Wikipedia, https://en.wikipedia.org/wiki/Famine; Shahla Shapouri and Stacey Rosen, "Fifty Years of U.S. Food Aid and Its Role in Reducing World Hunger," Economic Research Service, USDA, September 1, 2004, https://www.ers.usda.gov /amber-waves/2004/september/fifty-years-of-us-food-aid-and-its-role-in-reducing -world-hunger/.

34. UN Security Council, "Amid Humanitarian Funding Gap, 20 Million People Across Africa, Yemen at Risk of Starvation, Emergency Relief Chief Warns Security Council," UN Meetings Coverage and Press Releases, March 10, 2017, https:// www.un.org/press/en/2017/sc12748.doc.htm; Katrin Park, "The Great American Food Aid Boondoggle," *Foreign Policy,* December 10, 2019, https://foreignpolicy.com/2019 /12/10/america-wheat-hunger-great-food-aid-boondoggle/.

35. Sue Kirchhoff, "Surplus U.S. Food Supplies Dry Up," *ABC News,* May 3, 2008, https://abcnews.go.com/Business/story?id=4770135&page=1; IPCC, *Climate Change and Land: An IPCC Special Report on Climate Change, Desertification, Land Degradation, Sustainable Land Management, Food Security, and Greenhouse Gas Fluxes in Terrestrial Ecosystems* (2019), 358, sect. 5.2.2, https://www.ipcc.ch/srccl/; United Nations Convention to Combat Desertification, *National Report on Efforts to Mitigate Desertification in the Western United States: The First United States Report on Activities Relevant to the United Nations Convention to Combat Desertification,* 2006.

36. Robert Arnason, "Soil Erosion Costs Farmers $3.1 Billion a Year in Yield Loss: Scientist," *Western Producer,* January 31, 2019, https://www.producer.com/2019 /01/soil-erosion-costs-farmers-3-1-billion-a-year-in-yield-loss-scientist/; David A. Robinson et al., "On the Value of Soil Resources in the Context of Natural Capital and Ecosystem Service Delivery," *Soil Science Issues* 78 (2014): 685–700; IPCC, *Climate Change,* 56, 358.

Chapter 7. The Climate-Soil Duet

1. IPCC, *Climate Change and Land: An IPCC Special Report on Climate Change, Desertification, Land Degradation, Sustainable Land Management, Food Se-*

curity, and Greenhouse Gas Fluxes in Terrestrial Ecosystems (2019), https://www.ipcc
.ch/srccl/.

2. J. Blunden and D. S. Arndt, eds., *A Look at 2019: Takeaway Points from the State of the Climate* (Boston: Bulletin of the American Meteorological Society, 2020), https://www.ametsoc.org/index.cfm/ams/publications/bulletin-of-the-american-meteorological-society-bams/state-of-the-climate/.

3. IPCC, *Climate Change,* 11, 61.

4. "How Can Climate Change Affect Natural Disasters?," Climate and Land Use Change, USGS, https://www.usgs.gov/faqs/how-can-climate-change-affect-natural-disasters-1?qt-news_science_products=0#qt-news_science_products; Linlin Li et al., "A Modest 0.5-m Rise in Sea Level Will Double the Tsunami Hazard in Macau," *Science Advances* 4 (2018): eaat1180; Faith Ka Shun Chan et al., "Flood Risk in Asia's Urban Mega-Deltas: Drivers, Impacts and Response," *Environment and Urbanization ASIA* 3 (2012): 41–61.

5. Randy Schnepf, *U.S. International Food Aid Programs: Background and Issues,* CRS Report R41072 (2016), 12; Charles E. Hanrahan, *Indian Ocean Earthquake and Tsunamis: Food Aid Needs and the U.S. Response,* CRS Report RS22027 (2005), 2; Blunden and Arndt, *Look at 2019,* 1–11.

6. Senay Habtezion, "Gender and Climate Change" (New York: United Nations Development Programme, 2016), 5.

7. "5 Natural Disasters That Beg for Climate Action," Oxfam International, https://www.oxfam.org/en/5-natural-disasters-beg-climate-action; United Nations, *Climate Change and Indigenous Peoples,* 2007, https://www.un.org/en/events/indigenousday/pdf/Backgrounder_ClimateChange_FINAL.pdf.

8. Nora E. Torres Castillo et al., "Impact of Climate Change and Early Development of Coffee Rust: An Overview of Control Strategies to Preserve Organic Cultivars in Mexico," *Science of the Total Environment* 738 (2020): 140225.

9. Maximilian Heath and Ana Mano, "Argentina, Brazil Monitor Massive Locust Swarm; Crop Damage Seen Limited," Reuters, June 25, 2020, https://www.reuters.com/article/us-argentina-brazil-grains-locusts/argentina-brazil-monitor-massive-locust-swarm-crop-damage-seen-limited-idUSKBN23W34K; Mélissa Goden, "Swarms of Up to 80 Million Locusts Decimating Crops in East Africa, Threatening Food Security for 13 Million People," *Time,* February 14, 2020, https://time.com/5784323/un-locust-east-africa/; "FAO Welcomes Additional €15 Million from the European Union to Fight Desert Locusts and Their Impacts on Food Security," FAO, July 8, 2020, http://www.fao.org/news/story/en/item/1296770/icode/.

10. Muhammad Farooq et al., "Soil Degradation and Climate Change in South Asia," in *Soil and Climate,* ed. Rattan Lal and B. A. Stewart (New York: CRC Press, 2018), 330–332; Merritt R. Turetsky et al., "Global Vulnerability of Peatlands to Fire and Carbon Loss," *Nature Geoscience* 8 (2015): 11–14.

11. IPCC, *Climate Change,* 6, 11; Ottmar Edenhofer et al., eds., *Climate Change 2014: Mitigation of Climate Change; Contribution of Working Group III to the Fifth Assessment Report of the Intergovernmental Panel on Climate Change* (New York: Cambridge University Press, 2014); David A. N. Ussiri and Rattan Lal, *Carbon Sequestration for Climate Change Mitigation and Adaptation* (Cham, Switzerland: Springer International, 2017); Jonathan Sanderman, Tomislav Hengl, and Gregory J. Fiske, "Soil Carbon Debt of 12,000 Years of Human Land Use," *PNAS* 114 (2017): 9575–9580.

12. Joseph M. Prospero and Olga L. Mayol-Bracero, "Understanding the Transport and Impact of African Dust on the Caribbean Basin," *Bulletin of the American Meteorological Society* 94 (2003): 1329–1337; Pablo Méndez Lázaro, quoted in Sabrina Imbler, "A Giant Dust Storm Is Heading Across the Atlantic," *Atlantic,* June 24, 2020, https://www.theatlantic.com/science/archive/2020/06/saharan-dust-storms -giving-earth-life/613441/; Cornelius Oertel et al., "Greenhouse Gas Emissions from Soils: A Review," *Geochemistry* 76 (2016): 327–352.

13. Pete Smith et al., "Agriculture, Forestry and Other Land Use (AFOLU)," in Edenhofer et al., *Climate Change 2014,* 811–922.

14. Merritt R. Turetsky et al., "Global Vulnerability of Peatlands to Fire and Carbon Loss," *Nature Geoscience* 8 (2015): 11–14; Raymond R. Weil and Nyle C. Brady, *Nature and Properties of Soils,* 15th ed. (London: Pearson, 2017), 296.

15. Clifton Bain and Emma Goodyer, *Horticulture and Peatlands: A Discussion Briefing for Scotland's National Peatland Plan Steering Group* (IUCN UK Peatland Programme, 2016), 1–6; Martin Evans and John Lindsay, "The Impact of Gully Erosion on Carbon Sequestration in Blanket Peatlands," *Climate Research* 45 (2010): 31–41; Niall McNamara et al., "Gully Hotspot Contribution to Landscape Methane (CH_4) and Carbon Dioxide (CO_2) Fluxes in the Northern Peatland," *Science Total Environment* 404 (2008): 354–360; Richard Lindsay, Richard Birnie, and Jack Clough, *IUCN UK Committee Peatland Programme Briefing Note No. 9: Weathering, Erosion and Mass Movement of Blanket Bog* (University of East London, 2014), 1–6; Virginia Gewin, "How Peat Could Protect the Planet," *Nature* 578 (2020): 204–208.

16. World Wildlife Fund, "8 Things to Know about Palm Oil," WWF, January 17, 2020, https://www.wwf.org.uk/updates/8-things-to-know-about-palm-oil; Lulie

Melling et al., "Soil CO$_2$ Fluxes from Different Ages of Oil Palm in Tropical Peatland of Sarawak, Malaysia," in *Soil Carbon,* ed. Alfred E. Hartemink and Kevin McSweeney (New York: Springer, 2014), 447–455; Jordan Hanania et al., "Gigatonne," Energy Education, University of Calgary, https://energyeducation.ca/encyclopedia/Gigatonne.

17. Bowen Zhang et al., "Methane Emissions from Global Rice Fields: Magnitude, Spatiotemporal Patterns, and Environmental Controls," *Global Biogeochemical Cycles* 30 (2016): 1246–1263; Virender Kumar and Jagdish K. Ladha, "Direct Seeding of Rice: Recent Developments and Future Research Needs," *Advances in Agronomy* 111 (2011): 297–413; "Rice Productivity," Ricepedia, Research Program on Rice, http://ricepedia.org/rice-as-a-crop/rice-productivity; Kewei Yu and William H. Patrick Jr., "Redox Window with Minimum Global Warming Potential Contribution from Rice Soils," *Soil Science Society of America Journal* 68 (2004): 2086–2091.

18. Yu Jiang et al., "Higher Yields and Lower Methane Emissions with New Rice Cultivars," *Global Change Biology* 23 (2017): 4728–4738; Yu Jiang et al., "Acclimation of Methane Emissions from Rice Paddy Fields to Straw Addition," *Science Advances* 5 (2019): eaau9038; Yuanfeng Cai et al., "Conventional Methanotrophs Are Responsible for Atmospheric Methane Oxidation in Paddy Soils," *Nature Communications* 7 (June 2016): 11728.

19. Kimberly P. Wickland et al., "Effects of Permafrost Melting on CO$_2$ and CH$_4$ Exchange of a Poorly Drained Black Spruce Lowland," *Journal of Geophysical Research* 111 (2006): G02011; Blunden and Arndt, *Look at 2019,* 1–11.

20. A. R. Ravishankara, John S. Daniel, and Robert W. Portmann, "Nitrous Oxide (N$_2$O): The Dominant Ozone-Depleting Substance Emitted in the 21st Century," *Science* 326 (2009): 123–125; David B. Parker et al., "Enteric Nitrous Oxide Emissions from Beef Cattle," *Professional Animal Scientist* 34 (2018): 594–607.

21. IPCC, *Climate Change,* 11, 46.

22. Elizabeth A. Ainsworth and Stephen P. Long, "30 Years of Free-Air Carbon Dioxide Enrichment (FACE): What Have We Learned About Future Crop Productivity and Its Potential for Adaptation?," *Global Change Biology* 27 (2021): 27–49.

23. Y. Govaerts and A. Lattanzio, "Surface Albedo Response to Sahel Precipitation Changes," *Eos* 88 (2007): 25–26.

24. Philipp Mueller, *The Sahel Is Greening* (London: Global Warming Policy Foundation, 2011), 1–13; Lennart Olsson, "Greening of the Sahel," Encyclopedia of Earth, updated July 27, 2012, https://editors.eol.org/eoearth/wiki/Greening_of_the _Sahel; Lennart Olsson, L. Eklundh, and J. Ardö, "A Recent Greening of the

Sahel—Trends, Patterns and Potential Causes," *Journal of Arid Environments* 63 (2005): 556–566.

25. IPCC, *Climate Change,* 8, 44.

26. Rattan Lal, "Sequestering Carbon in Soils of Agro-Ecosystems," *Food Policy* 36 (2011): S33–S39; "How Can Climate Change Affect Natural Disasters?"

27. Dominic Woolf et al., "Biochar for Climate Mitigation: Navigating from Science to Evidence-Based Policy," in *Soil and Climate,* ed. Rattan Lal and B. A. Stewart, 220–248 (New York: CRC Press, 2018).

28. Turetsky, "Global Vulnerability," 11–14; Narayan Sastry, "Forest Fires, Air Pollution, and Mortality in Southeast Asia," *Demography* 39 (2002): 1–23.

29. Dennis Normile, "Parched Peatlands Fuel Indonesia's Blazes," *Science* 366 (2019): 18–19.

30. "The Paris Agreement," United Nations Climate Change, https://unfccc.int /process-and-meetings/the-paris-agreement/the-paris-agreement; William H. Schlesinger and Ronald Amundson, "Managing for Soil Carbon Sequestration: Let's Get Realistic," *Global Change Biology* 25 (2019): 386–389; Keith Paustian et al., "Climate-Smart Soils," *Nature* 532 (2016): 49–57.

31. Bijesh Maharjan, Saurav Das, and Bharat Sharma Acharya, "Soil Health Gap: A Concept to Establish a Benchmark for Soil Health Management," *Global Ecology and Conservation* 23 (2020): e01116; Paustian et al., "Climate-Smart Soils," 49–57; Ussiri and Lal, *Carbon Sequestration,* 327–341.

Chapter 8. Soil Stewards

1. David R. Montgomery, *Dirt: The Erosion of Civilizations, with a New Preface* (Berkeley: University of California Press, 2012).

2. Donald A. Davidson and Stephen P. Carter, "Micromorphological Evidence of Past Agricultural Practices in Cultivated Soils: The Impact of a Traditional Agricultural System on Soils in Papa Stour, Shetland," *Journal of Archaeological Science* 25 (1998): 827–838.

3. Manuel Arroyo-Kalin, "Amazonian Dark Earths: Geoarchaeology," in *Encyclopedia of Global Archaeology,* ed. Claire Smith (New York: Springer, 2014).

4. Michael E. Smith, *The Aztecs* (Hoboken, N.J.: John Wiley and Sons, 2013), table 3.1; Naomi Tomky, "Mexico's Famous Floating Gardens Return to Their Agricultural Roots," *Smithsonian Magazine,* January 31, 2017, https://www.smithsonian mag.com/travel/mexicos-floating-gardens-return-their-agricultural-roots-180961899/;

FAO, "Chinampas of Mexico City Were Recognized as an Agricultural Heritage System of Global Importance," http://www.fao.org/americas/noticias/ver/en/c/1118851/.

5. "Rice Terraces of the Philippine Cordilleras," UNESCO, https://whc .unesco.org/en/list/722/; Rogelio N. Concepcion, Edna Samar, and Mario Collado, *Multifunctionality of the Ifugao Rice Terraces in the Philippines* (Diliman, Quezon City, Philippines: Bureau of Soil and Water Management, 2006).

6. Christopher Poeplau and Axel Don, "Carbon Sequestration in Agricultural Soils via Cultivation of Cover Crops: A Meta-Analysis," *Agriculture, Ecosystems and Environment* 200 (2015): 33–41.

7. "6 Ways Indigenous Peoples Are Helping the World Achieve #ZeroHunger," FAO, September 8, 2017, http://www.fao.org/indigenous-peoples/news-article/en/c /1029002/.

8. A. Ford, "The Roots of the Maya Calendar," in *World History: Ancient and Medieval Eras,* ed. David Tipton et al. (online database, ABC-CLIO Solutions, 2012); Robert F. Spencer and Jesse D. Jennings, *The Native Americans* (New York: Harper and Row, 1977), 461–477.

9. Joost van Heerwaarden et al., "Genetic Signals of Origin, Spread, and Introgression in a Large Sample of Maize Landraces," *PNAS* 108 (2011): 1088–1092; Yoshihiro Matsuoka et al., "A Single Domestication for Maize Shown by Multilocus Microsatellite Genotyping," *PNAS* 99 (2002): 6080–6084.

10. Sheryl Luzzadder-Beach, Timothy P. Beach, and Nicholas P. Dunning, "Wetland Fields as Mirrors of Drought and the Maya Abandonment," *PNAS* 109 (2012): 3646–3651; David L. Lentz et al., "Molecular Genetic and Geochemical Assays Reveal Severe Contamination of Drinking Water Reservoirs at the Ancient Maya City of Tikal," *Scientific Reports* 10 (2020): 10316; Montgomery, *Dirt,* 74–78; Anabel Ford and Ronald Nigh, *The Maya Forest Garden: Eight Millennia of Sustainable Cultivation of the Woodlands* (New York: Routledge, 2015), 38, 77–96; Jared Diamond, *Collapse: How Societies Choose to Fail or Succeed* (London: Penguin, 2011), 159–160, 172–173.

11. Canadian Museum of History, "Maya Civilization," https://www.history museum.ca/cmc/exhibitions/civil/maya/mmc12eng.html; C. A. Petrie and J. Bates, "'Multi-Cropping,' Intercropping and Adaptation to Variable Environments in Indus South Asia," *Journal of World Prehistory* 30 (2017): 81–130; Anabel Ford and Ronald Nigh, "The Milpa Cycle and the Making of the Maya Forest Garden," *Research Reports in Belizean Archaeology* 7 (2010): 183–190; Daniel C. Allen, Bradley J. Cardinale,

and Theresa Wynn-Thompson, "Plant Biodiversity Effects in Reducing Fluvial Erosion Are Limited to Low Species Richness," *Ecology* 97 (2016): 17–24; Anabel Ford, "Maya Forest Garden," in *Encyclopedia of Global Archaeology*, ed. Claire Smith (Cham, Switzerland: Springer, 2018); Stewart A. W. Diemont et al., "Lacandon Maya Forest Management: Restoration of Soil Fertility Using Native Tree Species," *Ecological Engineering* 28 (2006): 205–212.

12. Ford and Nigh, *Maya Forest Garden,* 38, 41–68; David Webster, *The Fall of the Ancient Maya: Solving the Mystery of the Maya Collapse* (London: Thames and Hudson, 2002), 348; Michael D. Coe and Stephen Houston, *The Maya,* 9th ed. (London: Thames and Hudson, 2015), 231; Ellen Gray, "Landsat Top Ten—International Borders: Mexico and Guatemala," NASA, July 23, 2012, https://www.nasa.gov/mission _pages/landsat/news/40th-top10-mexico-guatemala.html; Tom Sever, "Archeological Research in the Petén, Guatemala," n.d., NASA, https://weather.msfc.nasa.gov /archeology/peten.html; Betsy Mason, "Landsat's Most Historically Significant Images of Earth from Space," July 23, 2012, Wired, https://www.wired.com/2012/07 /landsat-40-significant-images/.

13. Ronald Nigh and Stewart A. W. Diemont, "The Maya Milpa: Fire and the Legacy of Living Soil," *Frontiers in Ecology and the Environment* 11 (2013): e45–e54.

14. Montgomery, *Dirt,* 74–78; Laura C. Schneider, "Bracken Fern Invasion in Southern Yucatán: A Case for Land-Change Science," *Geographical Review* 94 (2004): 229–241.

15. Flavio S. Anselmetti et al., "Quantification of Soil Erosion Rates Related to Ancient Maya Deforestation," *Geology* 35 (2007): 915; Timothy Beach et al., "Impacts of the Ancient Maya on Soil Erosion in the Central Maya Lowlands," *Catena* 65 (2006): 166–178.

16. Scott Macrae and Gyles Iannone, "Understanding Ancient Maya Agricultural Terrace Systems Through LIDAR and Hydrological Mapping," *Advances in Archaeological Practice* 4 (2016): 371–392.

17. Ronald Nigh, "Trees, Fire, and Farmers: Making Woods and Soil in the Maya Forest," *Journal of Ethnobiology* 28 (2008): 231–243; Ford and Nigh, "Milpa Cycle," 183–190; Mark Stevenson, "Mexico's Indigenous Lacandon Battle Settlers over Rainforest," Associated Press, October 11, 2019, https://apnews.com/article /4b066fcf65ee494ab36c144904994725; Anabel Ford, Keith C. Clarke, and Gary Raines, "Modeling Settlement Patterns of the Late Classic Maya Civilization with Bayesian Methods and Geographic Information Systems," *Annals of the Association of American Geographers* 99 (2009): 496–520; Nigh and Diemont, "Maya Milpa."

18. Alisher Mirzabaev, Jiang Wu, et al., "Desertification," in IPCC, *Climate Change and Land: An IPCC Special Report on Climate Change, Desertification, Land Degradation, Sustainable Land Management, Food Security, and Greenhouse Gas Fluxes in Terrestrial Ecosystems* (2019), https://www.ipcc.ch/srccl/; J. A. Sandor et al., *Soil Knowledge Embodied in a Native American Runoff Agroecosystem* (Bangkok: World Congress of Soil Science, 2002).

19. David A. Cleveland et al., "Zuni Farming and United States Government Policy: The Politics of Biological and Cultural Diversity in Agriculture," *Agriculture and Human Values* 12 (1995): 2–18; Gary Paul Nabhan, Patrick Pynes, and Tony Joe, "Safeguarding Species, Languages, and Cultures in the Time of Diversity Loss: From the Colorado Plateau to Global Hotspots," *Annals of the Missouri Botanical Garden* 89 (2002): 164–175.

20. Jeffrey A. Homburg, Jonathan A. Sandor, and Jay B. Norton, "Anthropogenic Influences on Zuni Agricultural Soils," *Geoarchaeology* 20 (2005): 661–693.

21. Jonathan A. Sandor, "Biogeochemical Studies of a Native American Runoff Agroecosystem," *Geoarchaeology* 22 (2007): 359–386; Sandor, *Soil Knowledge;* Kelly M. Coburn, Edward R. Landa, and Gail E. Wagner, *Of Silt and Ancient Voices: Water and the Zuni Land and People* (Buffalo, N.Y.: National Center for Case Study Teaching, University of Buffalo, 2014).

22. Cleveland et al., "1995 Zuni Farming," 2–18.

23. Fanny Wonu Veys, *Mana Māori: The Power of New Zealand's First Inhabitants* (Leiden: Leiden University Press, 2010).

24. Jessica Hutchings, Jo Smith, and Garth Harmsworth, "Elevating the Mana of Soil Through the Hua Parakore Framework," *MAI Journal* 7, no. 1 (2018).

25. Garth R. Harmsworth and N. Roskruge, "Indigenous Māori Values, Perspectives and Knowledge of Soils in Aotearoa-New Zealand: Chapter 9—Beliefs, and Concepts of Soils, the Environment and Land," in *The Soil Underfoot: Infinite Possibilities for a Finite Resource,* ed. G. J. Churchman and E. R. Landa (Boca Raton, Fla.: CRC Press, 2014), 111–126.

26. D. Rhodes, "Rehabilitation of Deforested Steep Slopes on the East Coast of New Zealand's North Island," *Unasylva* 52 (2001): 21–29; Science Learning Hub, "Middens," https://www.sciencelearn.org.nz/resources/1460-middens; Tara A. Kniskern et al., "Sediment Accumulation Patterns and Fine-Scale Strata Formation on the Waiapu River Shelf, New Zealand," *Marine Geology* 270 (2010): 188–201.

27. Brad Japhe, "The Wild Story of Manuka, the World's Most Coveted Honey," AFAR, April 20, 2018, https://www.afar.com/magazine/the-wild-story-of-manuka-the

-worlds-most-coveted-honey; Matthew Johnston et al., "Antibacterial Activity of Manuka Honey and Its Components: An Overview," *AIMS Microbiology* 4 (2018): 655–664.

Chapter 9. Soil Heroes

1. Damien Houlahan, "Preparing for the Future of Almonds: The Next 10 Years," Olam, https://www.olamgroup.com/investors/investor-library/olam-insights /issue-1-forging-ahead-creating-secure-future-almonds-californian-agriculture/pre paring-future-almonds-next-10-years.html; Delicia Warren, "Global Almond Industry Has Projected CAGR of More Than 7% Through 2028," American Journal of Trans-portation, https://www.ajot.com/insights/full/ai-global-almond-industry-has-projected -cagr-of-more-than-7-through-2028.

2. FAO, *Small Family Farms Country Factsheet: Malawi* (Rome: FAO, 2018); George Rapsomanikis, *The Economic Lives of Smallholder Farmers: An Analysis Based on Household Data from Nine Countries* (Rome: FAO, 2015); FAO, *Small Family Farms Country Factsheet: Guatemala* (Rome: FAO, 2018).

3. S. M. Crispin et al., "The 2001 Mouth and Foot Disease Epidemic in the United Kingdom: Animal Welfare Perspectives," *Reviews of Science and Technology* 21 (2002): 877–883; World Bank, "Impacts of COVID-19 on Commodity Markets Heaviest on Energy Prices: Lower Oil Demand Likely to Persist Beyond 2021," news release no. 2021/047/EFI, October 22, 2020, https://www.worldbank.org/en/news /press-release/2020/10/22/impact-of-covid-19-on-commodity-markets-heaviest-on -energy-prices-lower-oil-demand-likely-to-persist-beyond-2021; Christian Elleby et al., "Impacts of the COVID-19 Pandemic on the Global Agricultural Markets," *En-vironmental and Resource Economics* 76 (2020): 1067–1079.

4. "Farmer Suicides: A Global Phenomenon," Brewhouse, Perspective, Pra-gati, May 6, 2015, http://pragati.nationalinterest.in/2015/05/farmer-suicides-a-global -phenomena/; Vishnu Padmanabhan and Pooja Danteadia, "The Geography of Farmer Suicides," Mint, January 16, 2020, https://www.livemint.com/news/india /the-geography-of-farmer-suicides-11579108457012.html; Dominic Merriott, "Fac-tors Associated with the Farmer Suicide Crisis in India," *Journal of Epidemiology and Global Health* 6 (2016): 217–227; Center for Human Rights and Global Justice, *Every Thirty Minutes: Farmer Suicides, Human Rights, and the Agrarian Crisis in India* (New York: NYU School of Law, 2011); Matt Perdue, "A Deeper Look at the CDC Findings on Farm Suicides," National Farmers Union, November 27, 2018, https:// nfu.org/2018/11/27/cdc-study-clarifies-data-on-farm-stress-2/; Cora Peterson et al.,

"Suicide Rates by Major Occupational Group: 17 States, 2012 and 2015," *Morbidity and Mortality Weekly Report* 67 (2018): 1254–1260.

5. Robert A. Hoppe, "Profit Margin Increases with Farm Size," in *Structure and Finances of U.S. Farms: Family Farm Report, 2014 Edition,* EIB-132, U.S. Department of Agriculture, Economic Research Service, December 2014.

6. James M. MacDonald, Penni Korb, and Robert A. Hoppe, "Farm Size and the Organization of U.S. Crop Farming," ERR-152, U.S. Department of Agriculture, Economic Research Service, August 2013.

7. Caroline Schneider, "Aldo Leopold and the Coon Valley Watershed Conservation Project," Certified Crop Adviser, https://www.certifiedcropadviser.org/science-news/aldo-leopold-and-coon-valley-watershed-conservation-project/; Gregory Hitch, "Lessons from Coon Valley: The Importance of Collaboration in Watershed Management," Aldo Leopold Foundation, July 23, 2015, https://www.aldoleopold.org/post/lessons-from-coon-valley-the-importance-of-collaboration-in-watershed-management/.

8. C. B. Johnson and W. C. Moldenhauer, "Effect of Chisel versus Moldboard Plowing on Soil Erosion by Water," *Soil Science Society of America Journal* 43 (1979): 177–179.

9. David R. Montgomery, "Soil Erosion and Agricultural Sustainability," *PNAS* 104 (2014): 13268–13272; Ronald E. Phillips et al., "No-Tillage Agriculture," *Science* 208 (1980): 1108–1113; Tiago Santos Telles, Bastiaan Philip Reydon, and Alexandre Gori Maia, "Effects of No-Tillage on Agricultural Land Values in Brazil," *Land Use Policy* 76 (2018): 124–129; A. Kassam, T. Friedrich, and R. Derpsch, "Global Spread of Conservation Agriculture," *International Journal of Environmental Studies* 76, no. 1 (2019).

10. Tyrone B. Hayes et al., "Hermaphroditic, Demasculinized Frogs After Exposure to the Herbicide Atrazine at Low Ecologically Relevant Doses," *PNAS* 99 (2002): 5476–5480; Tolga Cavas, "In Vivo Genotoxicity Evaluation of Atrazine and Atrazine-Based Herbicide on Fish *Carassius auratus* Using the Micronucleus Test and the Comet Assay," *Food and Chemical Toxicology* 49 (2011): 1431–1435; Mariana Cruz Delcorso et al., "Effects of Sublethal and Realistic Concentrations of the Commercial Herbicide Atrazine in Pacu (*Piaractus mesopotamicus*): Long-Term Exposure and Recovery Assays," *Vet World* 13 (2020): 147–159; Agency for Toxic Substances and Disease Registry, *Toxicological Profile for Atrazine* (Atlanta, Ga.: U.S. Department of Health and Human Services, Public Health Service), https://www.atsdr.cdc.gov/toxprofiles/tp153-c1-b.pdf.

11. Graham Brookes and Peter Barfoot, "Global Income and Production Impacts of Using GM Crop Technology, 1996–2013," *GM Crops and Food* 6 (2015): 13–46; Srinivasa Konduru, John Kruse, and Nicholas Kalaitzandonakes, "The Global Economic Impacts of Roundup Ready Soybeans," in *Genetics and Genomics of Soybean,* ed. Gary Stacey, 375–395 (New York: Springer, 2008); Phillip N. Johnson and Jason Blackshear, "Economic Analysis of Roundup Ready Versus Conventional Cotton Varieties in the Southern High Plains of Texas," *Texas Journal of Agriculture and Natural Resources* 17 (2004): 87–96.

12. "Recent Trends in GE Adoption," Economic Research Service, U.S. Department of Agriculture, last updated July 17, 2020, https://www.ers.usda.gov/data -products/adoption-of-genetically-engineered-crops-in-the-us/recent-trends-in-ge -adoption.aspx.

13. Graham Brookes, Farzad Taheripour, and Wallace E. Tyner, "The Contribution of Glyphosate to Agriculture and Potential Impact of Restrictions on Use at the Global Level," *GM Crops and Food* 8 (2017): 216–228; Ian Heap and Stephen O. Duke, "Overview of Glyphosate-Resistant Weeds Worldwide," *Pest Management Science* 74 (2018): 1040–1049.

14. Kathryn Z. Guyton et al., "Carcinogenicity of Tetrachlorvinphos, Parathion, Malathion, Diazinon, and Glyphosate," *Lancet Oncology* 16 (2015): 490–491.

15. Nakian Kim et al., "Do Cover Crops Benefit Soil Microbiome? A Meta-Analysis of Current Research," *Soil Biology and Biochemistry* 241 (2020): 107701.

16. James D. Plourde, Bryan C. Pijanowski, and Burak K. Pekin, "Evidence for Increased Monoculture Cropping in the Central United States," *Agriculture, Ecosystems, and Environment* 165 (2013): 50–59.

17. Luis Damiano and Jarad Niemi, *Quantification of the Impact of Prairie Strips on Grain Yield at the Neal Smith National Wildlife Refuge* (Ames: Iowa State University Department of Statistics, 2020); Javed Iqbal et al., "Denitrification and Nitrous Oxide Emissions in Annual Croplands, Perennial Grass Buffers, and Restored Perennial Grasslands," *Soil Science Society of America Journal* 79 (2015); Adam G. Dolezal et al., "Native Habitat Mitigates Feast-Famine Conditions Faced by Honey Bees in an Agricultural Landscape," *PNAS* 116 (2019): 25147–25155.

18. Telles, Reydon, and Maia, "Effects of No-Tillage," 124–129.

19. Craig Mackintosh, "Worldwide Permaculture Network: Project Type Descriptions," Permaculture News, Permaculture Research Institute, January 4, 2011, https://www.permaculturenews.org/2011/01/04/worldwide-permaculture-network -project-type-descriptions/.

20. "The Four Principles of Organic Agriculture," IFOAM Organics International, https://www.ifoam.bio/why-organic/shaping-agriculture/four-principles-organic.

21. "Global Organic Area Continues to Grow," Fresh Plaza, February 17, 2020, https://www.freshplaza.com/article/9189536/global-organic-area-continues-to -grow/; Helga Willer, "Organic Market Worldwide: Observed Trends in the Last Few Years," Bio Eco Actual, October 3, 2020, https://www.bioecoactual.com/en /2020/03/10/organic-market-worldwide-observed-trends-in-the-last-few-years/; "Global Organic Market: Export Opportunity Analysis," Global Marketing Associates, May 6, 2020, http://www.globalmarketing1.com/food-beverage/global-organic -market-export-opportunity-analysis/; Verena Seufert, Navin Ramankutty, and Jonathan A. Foley, "Comparing the Yields of Organic and Conventional Agriculture," *Nature* 485 (2012): 229–234.

22. Madelon Lohbeck et al., "Drivers of Farmer-Managed Natural Regeneration in the Sahel: Lessons for Restoration," *Scientific Reports* 10 (2020): 15038.

23. Duncan Gromko, "In Semi-Arid Africa, Farmers Are Transforming the 'Underground Forest' into Life-Giving Trees," Ensia, Institute on the Environment, February 11, 2020, https://ensia.com/features/in-semi-arid-africa-farmers-are-trans forming-the-underground-forest-into-life-giving-trees/; Joachim N. Binam et al., "Effects of Farmer Managed Natural Regeneration on Livelihoods in Semi-Arid West Africa," *Environmental Economics and Policy Studies* 17 (2015): 543–575; Peter Weston et al., "Farmer-Managed Natural Regeneration Enhances Rural Livelihoods in Dryland West Africa," *Environmental Management* 55 (2015): 1402–1417.

24. Lohbeck et al., "Drivers"; J. Bayala et al., "Regenerated Trees in Farmers' Fields Increase Soil Carbon Across the Sahel," *Agroforestry Systems* 94 (2020): 401–415.

25. Mimi Hillenbrand et al., "Impacts of Holistic Planned Grazing with Bison Compared to Continuous Grazing with Cattle in South Dakota Shortgrass Prairie," *Agriculture, Ecosystems, and Environment* 279 (2019): 156–168; Barry Estabrook, "Meet Allan Savory, the Pioneer of Regenerative Agriculture," *Successful Farming,* March 8, 2018, https://www.agriculture.com/livestock/cattle/meet-allan-savory-the -pioneer-of-regenerative-agriculture.

26. Paige L. Stanley et al., "Impacts of Soil Carbon Sequestration on Life Cycle Greenhouse Gas Emissions in Midwestern USA Beef Finishing Systems," *Agricultural Systems* 162 (2018): 249–258.

27. "Urban Farms," United Community Centers, https://ucceny.org/urban-farm/.

28. Richard Schiffman, "The City's Buried Treasure Isn't Under the Dirt. It Is

the Dirt," *New York Times,* July 25, 2018, https://www.nytimes.com/2018/07/25/ny region/the-citys-buried-treasure-isnt-under-the-dirt-it-is-the-dirt.html.

29. Miigle+, "The Rise of Urban Farming," *Medium,* May 25, 2019, https://medium.com/@Miigle/the-rise-of-urban-farming-cf894db51784; Liz Stinson, "World's Largest Rooftop Urban Farm to Open in Paris Next Year," Curbed, August 15, 2019, https://www.curbed.com/2019/8/15/20806540/paris-rooftop-urban-farm-opening; Kimberly Lim and Kalpana Sunder, "From Singapore to India, Urban Farms Sprout Up as Coronavirus Leaves Bollywood Celebrities with Thyme on Their Hands," *South China Morning Post,* August 2, 2020, https://www.scmp.com/week-asia/people/article/3095592/singapore-india-urban-farms-sprout-coronavirus-leaves-bollywood.

Chapter 10. A World With Soil

1. "United Nations Environment Programme: Nairobi Declaration on the State of Worldwide Environment," *International Legal Materials* 21 (1982): 677; FAO and ITPS, *Status of the World's Soil Resources: Main Report* (Rome: FAO, 2015), 225; "Protocol on the Implementation of the Alpine Convention of 1991 in the Field of Soil Conservation: Soil Conservation Protocol," *Official Journal of the European Union* (December 2015).

2. Samantha Harrington, "How Climate Change Affects Mental Health," Yale Climate Connections, Yale Center for Environmental Communication, February 4, 2020, https://yaleclimateconnections.org/2020/02/how-climate-change-affects-mental-health/; Kari Marie Norgaard, "Cognitive and Behavioral Challenges in Responding to Climate Change" (working paper, The World Bank, Washington, D.C., May 2009).

3. Scott Barrett, *Environment and Statecraft: The Strategy of Environmental Treaty Making* (Oxford: Oxford University Press, 2003), 1–18; 4 Per 1000 Initiative (website), 4 per 1000, https://www.4p1000.org.

4. Budiman Minasny et al., "Soil Carbon 4 per Mille," *Geoderma* 292 (April 2017): 59–86; Rattan Lal, "Digging Deeper: A Holistic Perspective of Factors Affecting Soil Organic Carbon Sequestration in Agroecosystems," *Global Change Biology* 24 (2018): 3285–3301; Adam Chambers, Rattan Lal, and Keith Paustian, "Soil Carbon Sequestration Potential of US Croplands and Grasslands: Implementing the 4 per Thousand Initiative," *Journal Soil Water Conservation* 71 (2016): 68A–74A; William H. Schlesinger and Ronald Amundson, "Managing for Soil Carbon Sequestration: Let's Get Realistic," *Global Change Biology* 25 (2019): 386–389.

5. Schlesinger and Amundson, "Managing"; Bijesh Maharjan, Saurav Das, and

Bharat Sharma Acharya, "Soil Health Gap: A Concept to Establish a Benchmark for Soil Health Management," *Global Ecology and Conservation* 23 (2020): e01116.

6. Schlesinger and Amundson, "Managing," 386–389; Minasny et al., "Soil"; "The Paris Agreement," United Nations Framework Convention on Climate Change, https://unfccc.int/process-and-meetings/the-paris-agreement/the-paris-agreement.

7. R. A. Houghton, "The Annual Net Flux of Carbon to the Atmosphere from Changes in Land Use, 1850–1990," *Tellus B: Chemical and Physical Meteorology* 51 (1999): 298–313.

8. Chambers, Lal, and Paustian, "Soil Carbon Sequestration Potential." "Average American Carbon Footprint," Inspire, July 21, 2020, https://www.inspireclean energy.com/blog/clean-energy-101/average-american-carbon-footprint.

9. *The State and Future of U.S. Soils: Framework for a Federal Strategic Plan for Soil Science,* Subcommittee on Ecological Systems, Committee on Environment, Natural Resources, and Sustainability of the NSTC (December 2016).

10. Ed Maixner and Philip Brasher, "Carbon Markets Lure Farmers, but Will Benefits Be Enough to Hook Them?," Agri-Pulse, November 23, 2020, https://www.agri-pulse.com/articles/14880-carbon-markets-lure-farmers-but-are-benefits-enough-to-hook-them.

11. "Global Meat Production, 1961 to 2018," Our World in Data, https://our worldindata.org/grapher/global-meat-production; Mimi Hillenbrand et al., "Impacts of Holistic Planned Grazing with Bison Compared to Continuous Grazing with Cattle in South Dakota Shortgrass Prairie," *Agriculture, Ecosystems, and Environment* 279 (2019): 156–168.

12. Cass R. Sunstein, *How Change Happens* (Cambridge, Mass.: MIT Press, 2019); Malcolm Gladwell, *The Tipping Point: How Little Things Can Make a Big Difference* (Boston: Little, Brown, 2000).

13. Rachel Carson, *Silent Spring* (Boston: Houghton Mifflin, 1962); Mark Kitchell, director, "Evolution of Organic," April 20, 2017, https://evolutionoforganic.com.

14. Franklin Fearing, "Influence of the Movies on Attitudes and Behavior," *Annals of the American Academy of Political and Social Science* 254 (1947): 70–79; Marty Kaplan, "Thank You, Norman Lear," *Norman Lear Center* (blog), https://learcenter.org/thank-you-norman-lear/; William DeJong and Jay A. Winsten, "The Use of Mass Media in Substance Abuse Prevention," *Health Affairs* 9 (1990): 30–46; Deborah Glik et al., "Health Education Goes Hollywood: Working with Prime-Time and Daytime Entertainment Television for Immunization Promotion," *Journal of Health*

Communication 3 (2010): 263–282; Environment Media Association (website), https://www.green4ema.org; Jay A. Winsten, "Promoting Designated Drivers: The Harvard Alcohol Project," *American Journal of Preventative Medicine* 10 (1994): 11–14.

15. Anthony A. Leiserowitz, "Day After Tomorrow: Study of Climate Change Risk Perception," *Environment* 46 (2004): 23–37; Ron Von Burg, "Decades Away or the Day After Tomorrow?: Rhetoric, Film, and the Global Warming Debate, Critical Studies in Media," *Critical Studies in Media Communication* 29 (2012): 7–26; Bridie McGreavy and Laura Lindenfeld, "Entertaining Our Way to Engagement? Climate Change Films and Sustainable Development Values," *International Journal of Sustainable Development* 17 (2014): 123–136.

16. "More People Are Gaming in the U.S., and They're Doing So Across More Platforms," NPD, July 20, 2020, https://www.npd.com/wps/portal/npd/us/news/press-releases/2020/more-people-are-gaming-in-the-us/; J. Clement, "Number of Active Video Gamers Worldwide from 2015 to 2023," Statista, January 29, 2021, https://www.statista.com/statistics/748044/number-video-gamers-world/; Max Mastro, "Over 3 Billion People Play Video Games, New Report Reveals," Screen Rant, August 16, 2020, https://screenrant.com/how-many-people-play-video-games-dfc-2020/; Peter Moore, "Poll Results: Reading," YouGov, September 30, 2013, https://today.yougov.com/topics/arts/articles-reports/2013/09/30/poll-results-reading.

17. National Research Council, *Climate Intervention: Carbon Dioxide Removal and Reliable Sequestration* (Washington, D.C.: National Academies Press, 2015), 107; David Emerson, "Biogenic Iron Dust: A Novel Approach to Ocean Iron Fertilization as a Means of Large Scale Removal of Carbon Dioxide from the Atmosphere," *Frontiers in Marine Science* 6 (February 2019).

Selected Bibliography

Prologue

Amundson, Ronald, Asmeret Asefaw Berhe, Jan W. Hopmans, Carolyn Olson, A. Ester Sztein, and Donald L. Sparks. "Soil and Human Security in the 21st Century." *Science* 348 (2015): 1261071.

Cruse, Richard, D. Flanagan, J. Frankenberger, B. Gelder, D. Herzmann, D. James, W. Krajewski, et al. "Daily Estimates of Rainfall, Water Runoff, and Soil Erosion in Iowa." *Journal of Soil and Water Conservation* 61 (2006): 191.

FAO and ITPS. *Status of the World's Soil Resources: Main Report.* Rome: FAO, 2015.

Montgomery, David R. "Soil Erosion and Agricultural Sustainability." *PNAS* 104 (2014): 13268–13272.

Chapter 1. Dawning—An Invisible Crisis

Ussiri, David A. N., and Rattan Lal. *Carbon Sequestration for Climate Change Mitigation and Adaptation.* Cham, Switzerland: Springer International, 2017.

Chapter 2. Earth's Dark Matter

Djokic, Tara, Martin J. Van Kranendonk, Kathleen A. Campbell, Malcolm R. Walter, and Colin R. Ward. "Earliest Signs of Life on Land Preserved in ca. 3.5 Ga Hot Spring Deposits." *Nature Communications* 8 (2017): 15263.

Fierer, Noah. "Earthworms' Place on Earth." *Science* 366 (2019): 425–426.

Flemming, Hans-Curt, and Stefan Wuertz. "Bacteria and Archaea on Earth and Their Abundance in Biofilms." *Nature Reviews Microbiology* 17 (2019): 247–260.

Hütsch, Birgit W., Jürgen Augustin, and Wolfgang Merbach. "Plant Rhizo-deposition: An Important Source for Carbon Turnover in Soils." *Journal of Plant Nutrition and Soil Science* 165 (2002): 397–407.

Kumar, Rajeew, Sharad Pandey, and Apury Pandey. "Plant Roots and Carbon Sequestration." *Current Science* 91 (2006): 885–890.

Lambers, Hans. "Growth, Respiration, Exudation and Symbiotic Associations: The Fate of Carbon Translocated to the Roots." In *Root Development and Function,* edited by P. J. Gregory, J. V. Lake, and D. A. Rose, 125–145. Cambridge: Cambridge University Press, 1987.

Nguyen, Christophe. "Rhizodeposition of Organic C by Plants: Mechanisms and Controls." *Agronomy* 23 (2003): 375–396.

Tashiro, Takayuki, Akizumi Ishida, Masako Hori, Motoko Igisu, Mizuho Koike, Pauline Méjean, Naoto Takahata, Yuji Sano, and Tsuyoshi Komiya. "Early Trace of Life from 3.95 Ga Sedimentary Rocks in Labrador, Canada." *Nature* 549 (2017): 516–518.

Valley, John W. "A Cool Early Earth?" *Scientific American* 293 (2005): 58–63.

Wilde, Simon A., John W. Valley, William H. Peck, and Colin M. Graham. "Evidence from Detrital Zircons for the Existence of Continental Crust and Oceans on the Earth 4.4 Gyr Ago." *Nature* 409 (2001): 175–178.

Chapter 3. Earth Works

Baltz, Richard H. "Marcel Faber Roundtable: Is Our Antibiotic Pipeline Unproductive Because of Starvation, Constipation or Lack of Inspiration?" *Journal of Industrial and Microbial Biotechnology* 33 (July 2006): 507–513.

Costa, Ohana Y. A., Jos M. Raaijmakers, and Eiko E. Kuramae. "Microbial Extracellular Polymeric Substances: Ecological Function and Impact on Soil Aggregation." *Frontiers in Microbiology* 9 (July 2018): 1636.

Feller, Christian, Lydie Chapuis-Lardy, and Fiorenzo Ugolini. "The Representation of Soil in the Western Art: From Genesis to Pedogenesis." In *Soil and Culture,* edited by Edward R. Landa and Christian Feller, 3–22. Dordrecht: Springer Netherlands, 2009.

Hütsch, Birgit W., Jürgen Augustin, and Wolfgang Merbach. "Plant Rhizo-deposition: An Important Source for Carbon Turnover in Soils." *Journal of Plant Nutrition and Soil Science* 165 (2002): 397–407.

Jones, Martin K., and Xinyi Liu. "Origins of Agriculture in East Asia." *Science* 324 (2009): 730–731.

National Research Council. *The New Science of Metagenomics: Revealing the Secrets of Our Microbial Planet.* Washington, D.C.: National Academies Press, 2007.

Stewart, W. M., D. W. Dibb, A. E. Johnston, and T. J. Smyth. "The Contribution of Commercial Fertilizer Nutrients to Food Production." *Agronomy* 97 (2005): 1–6.

Tauger, Mark B. "The Origins of Agriculture and the Dual Subordination." In *Agriculture in World History,* 3–14. London: Routledge, 2010.

Torsvik, Vigdis, and Lise Øvreås. "Microbial Diversity and Function in Soil: From Genes to Ecosystems." *Current Opinion in Microbiology* 5 (2002): 240.

The United Nations World Water Development Report: Water for a Sustainable World: Facts and Figures. Paris: UNESCO, 2015.

Wang, B., and Y.-L. Qiu. "Phylogenetic Distribution and Evolution of Mycorrhizas in Land Plants." *Mycorrhiza* 16 (2006): 299–363.

Chapter 4. Chaos to Orders

Buol, Stanley W., Randal J. Southard, Robert C. Graham, and Paul A. McDaniel. *Soil Genesis and Classification.* 5th ed. Ames: Iowa State University Press, 2003.

Deckers, Jozef, Paul Driessen, Freddy Nachtergaele, and Otto Spaargaren. *World Reference Base for Soil Resources—in a Nutshell.* European Soil Bureau, European Soil Bureau, Research Report no. 7, January 2001.

Hans, Jenny. *Factors of Soil Formation: A System of Quantitative Pedology.* New York: McGraw-Hill, 1941.

Krasilnikov, Pavel, Juan-José Ibáñez Martí, Richard Arnold, and Serghei Shoba, eds. *A Handbook of Soil Terminology, Correlation and Classification.* London: Routledge, 2009.

Wallace, Robert S. "Record-Holding Plants." *Plant Sciences,* Encyclopedia

.com, updated December 30, 2020. https://www.encyclopedia.com/science/news-wires-white-papers-and-books/record-holding-plants.

West, L. T., M. J. Singer, and A. E. Hartemink, eds. "Introduction." In *The Soils of the USA*, 1–7. Cham, Switzerland: Springer, 2017.

Chapter 5. Wind, Water, and Plows

Arsenault, Chris. "Only 60 Years of Farming Left If Soil Degradation Continues." *Scientific American*, December 5, 2014. https://www.scientificamerican.com/article/only-60-years-of-farming-left-if-soil-degradation-continues/.

Chen, Xiaochao, Jie Zhang, Yanling Chen, Qian Li, Fanjun Chen, Lixing Yuan, and Guohua Mi. "Changes in Root Size and Distribution in Relation to Nitrogen Accumulation During Maize Breeding in China." *Plant Soil* 374 (2014): 121–130.

Cox, Craig, Andrew Hug, and Nils Bruzelius. *Losing Ground*. Washington, D.C.: Environmental Working Group, April 2011.

Daniels, R. B. "Soil Erosion and Degradation in the Southern Piedmont of the USA." In *Land Transformation in Agriculture*, edited by M. G. Wolman and F. G. A. Fournier, 407–428. New York: John Wiley and Sons, 1987.

den Biggelaar, Christoffel, Rattan Lal, Hari Eswaran, Vincent E. Breneman, and Paul F. Reich. "Crop Yield Losses to Soil Erosion at Regional and Global Scales: Evidence from Plot-Level and GIS Data." In *Land Quality, Agricultural Productivity, and Food Security: Biophysical Processes and Economic Choices at Local, Regional, and Global Levels*, edited by Keith Wiebe, 262–279. Cheltenham, UK: Edward Elgar, 2003.

Egan, Timothy. *The Worst Hard Time: The Untold Story of Those Who Survived the Great American Dust Bowl*. New York: Houghton Mifflin Harcourt, 2006.

Gelder, Brian, Tim Sklenar, David James, Daryl Herzmann, Richard Cruse, Karl Gesch, and John Laflen. "The Daily Erosion Project: Daily Estimates of Water Runoff, Soil Detachment, and Erosion." *Earth Surface Processes and Landforms* 43 (2018): 1105–1117.

Hamza, Waleed. "The Nile Delta." In *The Nile,* edited by H. J. Dumont, 75–94. Dordrecht: Springer Netherlands, 2009.

Hsieh, Y. P., K. T. Grant, and G. C. Bugna. "A Field Method for Soil Erosion Measurements in Agricultural and Natural Lands." *Journal of Soil and Water Conservation* 64 (2009): 374.

IPCC. *Climate Change and Land: An IPCC Special Report on Climate Change, Desertification, Land Degradation, Sustainable Land Management, Food Security, and Greenhouse Gas Fluxes in Terrestrial Ecosystems.* 2019. https:// www.ipcc.ch/srccl/.

Jarchow, Meghann, E., and Matt Liebman. "Tradeoffs in Biomass and Nutrient Allocation in Prairies and Corn Managed for Bioenergy Production." *Crop Science* 52 (2012): 1330–1342.

Jefferson, Thomas. *Thomas Jefferson's Garden Book.* Edited by E. M. Betts. Monticello, Va.: Thomas Jefferson Foundation, 1999.

Jelinski, Nicolas A., Benjamin Campforts, Jane A. Willenbring, Thomas E. Schumacher, Sheng Li, David A. Lobb, Sharon K. Papiernik, and Kyungsoo Yoo. "Meteoric Beryllium-10 as a Tracer of Erosion Due to Postsettlement Land Use in West-Central Minnesota, USA." *Journal of Geophysical Research: Earth Surface* 124 (2019): 874–901.

King, C., N. Baghdadi, V. Lecomte, and O. Cerdan. "The Application of Remote-Sensing Data to Monitoring and Modelling of Soil Erosion." *Catena* 62 (2005): 79–93.

Laflen, John M., and Dennis C. Flanagan. "The Development of U.S. Soil Erosion Prediction and Modeling." *International Soil and Water Conservation Research* 1 (2013): 2.

Lal, Rattan, and William C. Moldenhauer. "Effects of Soil Erosion on Crop Productivity." *Critical Reviews in Plant Sciences* 5 (1987): 303–367.

Montgomery, David R. "Soil Erosion and Agricultural Sustainability." *PNAS* 104 (2014): 13268–13272.

Munne-Bosch, Sergi. "Perennial Roots to Immortality." *Plant Physiology* 166 (2014): 720–725.

Portenga, Eric W., and Paul R. Bierman. "Understanding Earth's Eroding Surface with [10]Be." *Geological Society of America Today* 21 (2011): 4–10.

Porto, Paolo, Des E. Walling, Vito Ferro, and Costanza di Sefano. "Validating Erosion Rate Estimates Provided by Caesium-137 Measurements for Two Small Forested Catchments in Calabria, Southern Italy." *Land Degradation and Development* 14 (2007): 389–408.

Ravi, Sujith, Paolo D'Odorico, David D. Breshears, Jason P. Field, Andrew S. Goudie, Travis E. Huxman, Junran Li, et al. "Aeolian Processes and the Biosphere." *Reviews of Geophysics* 49 (2011): 1.

Tian, Qiuying, Fanjun Chen, Fusuo Zhang, and Guohua Mi. "Genotypic Difference in Nitrogen Acquisition Ability in Maize Plants Is Related to the Coordination of Leaf and Root Growth." *Journal of Plant Nutrition* 29 (2006): 317–330.

Veenstra, Jessica J., and C. Lee Burras. "Soil Profile Transformation After 50 Years of Agricultural Land Use." *Soil Science Society of America Journal* 79 (2015): 1154–1162.

Wilkinson, Bruce H., and Brandon J. McElroy. "The Impact of Humans on Continental Erosion and Sedimentation." *Geological Society of America Bulletin* 119 (2007): 140–156.

Chapter 6. Rocky Planet

Almagro, André, Paulo Tarso S. Oliveira, Mark A. Mearing, and Stefan Hagemann. "Projected Climate Change Impacts in Rainfall Erosivity over Brazil." *Scientific Reports* 7 (2017): 8130.

"Bhutan: Committed to Conservation." World Wildlife Foundation. https://www.worldwildlife.org/projects/bhutan-committed-to-conservation.

Borrelli, Pasquale, David A. Robinson, Larissa R. Fleischer, Emanuele Lugato, Cristiano Ballabio, Christine Alewell, Katrin Meusburger, et al. "An Assessment of the Global Impact of 21st Century Land Use Change on Soil Erosion." *Nature Communications* 8 (2017): 2013.

Cruse, Richard M. *Economic Impacts of Soil Erosion in Iowa.* Leopold Center Completed Grant Reports, 2016.

FAO. *Small Family Farms Country Factsheet: Indonesia.* Rome: FAO, 2018.

Foley, Jonathan A., Navin Ramankutty, Kate A. Brauman, Emily S. Cassidy, James S. Gerber, Matt Johnston, Nathanial D. Mueller, et al. "Solutions for a Cultivated Planet." *Nature* 478 (2011): 337–342.

Franko, Ivan. "Chernozems of Ukraine: Past, Present, and Future Perspectives." *Soil Science Annual* 70 (2019): 193–197.

Khor, Martin. "Land Degradation Causes $10 Billion Loss to South Asia Annually." *Global Policy Forum.* https://www.globalpolicy.org/global-taxes/49705-land-degradation-causes-10-billion-loss-to-south-asi.

Kinealy, Christine. "Saving the Irish Poor: Charity and the Great Famine." In *The 1846–1851 Famine in Ireland: Echoes and Repercussions,* Cahiers du MIMMOC, December 2015. https://doi.org/10.4000/mimmoc.1845.

Liang, Yanru, Rattan Lal, Shengli Guo, Ruiqiang Liu, and Yaxian Hu. "Impacts of Simulated Erosion and Soil Amendments on Greenhouse Gas Fluxes and Maize Yield in Miamian Soil of Central Ohio." *Scientific Reports* 8 (2018): 520.

Liu, Xiobang, Charles Lee Burras, Yuri S. Kravchenko, Artigas Duran, Ted Huffman, Hector Morras, Guillermo Studdert, Xingyi Zhang, Richard M. Cruse, and Xiaohui Yuan. "Overview of Mollisols in the World: Distribution, Land Use and Management." *Canadian Journal of Soil Science* 92 (2011): 383–402.

Milesi, Orlando, and Marianela Jarroud. "Soil Degradation Threatens Nutrition in Latin America." *Inter Press Service,* June 15, 2016. http://www.ipsnews.net/2016/06/soil-degradation-threatens-nutrition-in-latin-america/.

Nkonya, Ephraim, Weston Anderson, Edward Kato, Jawoo Koo, Alisher Mirzabaev, Joachim von Braun, and Stefan Meyer. "Global Cost of Land Degradation." In *Economics of Land Degradation and Improvement: A Global Assessment for Sustainable Development,* edited by Ephraim Nkonya, Alisher Mirzabaev, and Joachim von Braun, 117–165. Cham, Switzerland: Springer International, 2016.

Pimentel, David, C. Harvey, P. Resosudarmo, K. Sinclair, D. Kurz, M. McNair, S. Crist, et al. "Environmental and Economic Costs of Soil Erosion and Conservation Benefits." *Science* 267 (1995): 1120.

PwC Brazil. *Agribusiness in Brazil: An Overview.* 2013. https://www.pwc.com.br/pt/publicacoes/setores-atividade/assets/agribusiness/2013/pwc-agribusiness-brazil-overview-13.pdf.

Rabotyagov, Sergey S., Todd D. Campbell, Michael White, Jeffrey G. Arnold,

Jay Atwood, M. Lee Norfleet, Catherine L. Kling, et al. "Cost-Effective Targeting of Conservation Investments to Reduce the Northern Gulf of Mexico Hypoxic Zone." *PNAS* III (2014): 18530–18535.

Ray, Deepak K., Navin Ramankutty, Nathaniel D. Mueller, Paul C. West, and Johnathan A. Foley. "Recent Patterns of Crop Yield Growth and Stagnation." *Nature Communications* 3 (2012): 1293.

Repetto, Robert, "Soil Loss and Population Pressure on Java." *AMBIO: A Journal of the Human Environment* 15 (1986): 14–18.

Robinson, David A., I. Fraser, E. J. Dominati, B. Davíðsdóttir, J. O. G. Jónsson, L. Jones, S. B. Jones, et al. "On the Value of Soil Resources in the Context of Natural Capital and Ecosystem Service Delivery," *Soil Science Issues* 78 (2014): 685–700.

Royal Government of Bhutan. *Bhutan: In Pursuit of Sustainable Development.* National Report for the United Nations Conference on Sustainable Development, 2012. https://sustainabledevelopment.un.org/content/documents/798bhutanreport.pdf.

Rudiarto, Iwan, and W. Doppler. "Impact of Land Use Change in Accelerating Soil Erosion in Indonesian Upland Area: A Case of Dieng Plateau, Central Java—Indonesia." *International Journal of AgriScience* 3 (2013): 574.

Sarkar, Dipak, Abul Kalam Azad, S. K. Sing, and Nasrin Akter, eds. *Strategies for Arresting Land Degradation in South Asian Countries.* Dhaka: SAARC Agriculture Centre, 2011.

Snyder, Timothy. *Black Earth: The Holocaust as History and Warning.* New York: Tim Duggan Books, 2016.

Telles, Tiago Santos, Sonia Carmela Falci Dechen, Luiz Gustavo Antonio de Souza, and Maria de Fátima Guimarães. "Valuation and Assessment of Soil Erosion Costs." *Scientia Agricola* 70 (2013): 209–216.

UN Security Council. "Amid Humanitarian Funding Gap, 20 Million People Across Africa, Yemen at Risk of Starvation, Emergency Relief Chief Warns Security Council." UN Meetings Coverage and Press Releases, March 10, 2017. https://www.un.org/press/en/2017/sc12748.doc.htm

Verma, Ritu. *Gender, Land, and Livelihoods in East Africa: Through Farmers' Eyes.* Ottawa, Ont.: International Development Research Centre, 2001.

Chapter 7. The Climate-Soil Duet

Evans, Martin, and John Lindsay. "The Impact of Gully Erosion on Carbon Sequestration in Blanket Peatlands." *Climate Research* 45 (2010): 31–41.

Gewin, Virginia. "How Peat Could Protect the Planet." *Nature* 578 (2020): 204–208.

IPCC. *Climate Change and Land: An IPCC Special Report on Climate Change, Desertification, Land Degradation, Sustainable Land Management, Food Security, and Greenhouse Gas Fluxes in Terrestrial Ecosystems.* 2019. https://www.ipcc.ch/srccl/.

Jiang, Yu, Kees Jan van Groenigen, Shan Huang, Bruce A. Hungate, Chris van Kessel, Shuijin Hu, Jun Zhang, et al. "Higher Yields and Lower Methane Emissions with New Rice Cultivars." *Global Change Biology* 23 (2017): 4728–4738.

Melling, Lulie, Kah Joo Goh, Auldry Chaddy, and Ryusuke Hatano. "Soil CO_2 Fluxes from Different Ages of Oil Palm in Tropical Peatland of Sarawak, Malaysia." In *Soil Carbon,* edited by Alfred E. Hartemink and Kevin McSweeney, 447–455. New York: Springer, 2014.

Oertel, Cornelius, Jörg Matschullat, Kamal Zurba, Frank Zimmermann, and Stefan Erasmi. "Greenhouse Gas Emissions from Soils: A Review." *Geochemistry* 76 (2016): 327–352.

Olsson, Lennart, L. Eklundh, and J. Ardö. "A Recent Greening of the Sahel: Trends, Patterns and Potential Causes." *Journal of Arid Environments* 63 (November 2005): 556–566.

Paustian, Keith, Johannes Lehmann, Stephen Ogle, David Reay, G. Philip Robertson, and Pete Smith. "Climate-Smart Soils." *Nature* 532 (2016): 49–57.

Ravishankara, A. R., John S. Daniel, and Robert W. Portmann. "Nitrous Oxide (N_2O): The Dominant Ozone-Depleting Substance Emitted in the 21st Century." *Science* 326 (2009): 123–125.

Turetsky, Merritt R., Brian Benscoter, Susan Page, Guillermo Rein, Guido R. van der Werf, and Adam Watts. "Global Vulnerability of Peatlands to Fire and Carbon Loss." *Nature Geoscience* 8 (2015): 11–14.

United Nations. *Climate Change and Indigenous Peoples,* 2007. https://www.un.org/en/events/indigenousday/pdf/Backgrounder_ClimateChange_FINAL.pdf.

Ussiri, David A. N., and Rattan Lal. *Carbon Sequestration for Climate Change Mitigation and Adaptation.* Cham, Switzerland: Springer International, 2017.

Woolf, Dominic, Johannes Lehmann, Annette Cowie, Maria Luz Cayuela, Thea Whitman, and Saran Sohi. "Biochar for Climate Mitigation: Navigating from Science to Evidence-Based Policy." In *Soil and Climate,* edited by Rattan Lal and B. A. Stewart, 219–248. New York: CRC Press, 2018.

Zhang, Bowen, Hangin Tian, Wei Ren, Bo Tao, Chaoqun Lu, Jia Yang, Kamaljit Banger, and Shufen Pan. "Methane Emissions from Global Rice Fields: Magnitude, Spatiotemporal Patterns, and Environmental Controls." *Global Biogeochemical Cycles* 30 (2016): 1246–1263.

Chapter 8. Soil Stewards

Beach, Timothy, N. Dunning, S. Luzzadder-Beach, D. E. Cook, and J. Lohse. "Impacts of the Ancient Maya on Soil Erosion in the Central Maya Lowlands." *Catena* 65 (2006): 166–178.

Cleveland, David A., Fred Bowannie Jr., Donald F. Eriacho, Andrew Laahty, and Eric Perramond. "Zuni Farming and United States Government Policy: The Politics of Biological and Cultural Diversity in Agriculture." *Agriculture and Human Values* 12 (1995): 2–18.

Ford, Anabel, and Ronald Nigh. "The Milpa Cycle and the Making of the Maya Forest Garden." *Research Reports in Belizean Archaeology* 7 (2010): 183–190.

Harmsworth, Garth R., and N. Roskruge. "Indigenous Māori Values, Perspectives and Knowledge of Soils in Aotearoa-New Zealand: Chapter 9—Beliefs, and Concepts of Soils, the Environment and Land." In *The Soil Underfoot: Infinite Possibilities for a Finite Resource,* edited by G. J. Churchman and E. R. Landa, 111–126. Boca Raton, Fla.: CRC Press, 2014.

Japhe, Brad. "The Wild Story of Manuka, the World's Most Coveted Honey." AFAR, April 20, 2018. https://www.afar.com/magazine/the-wild-story -of-manuka-the-worlds-most-coveted-honey.

Lentz, David L., Trinity L. Hamilton, Nicholas P. Dunning, Vernon L. Scarborough, Todd P. Luxton, Anne Vonderheide, Eric J. Tepe, et al. "Molecular Genetic and Geochemical Assays Reveal Severe Contamination of Drinking Water Reservoirs at the Ancient Maya City of Tikal." *Scientific Reports* 10 (2020): 10316.

Matsuoka, Yoshihiro, Yves Vigouroux, Major M. Goodman, Jesus Sanchez G., Edward Buckler, and John Doebley. "A Single Domestication for Maize Shown by Multilocus Microsatellite Genotyping." *PNAS* 99 (2002): 6080–6084.

Montgomery, David R. *Dirt: The Erosion of Civilizations, with a New Preface.* Berkeley: University of California Press, 2012.

Poeplau, Christopher, and Axel Don. "Carbon Sequestration in Agricultural Soils via Cultivation of Cover Crops: A Meta-Analysis." *Agriculture, Ecosystems and Environment* 200 (2015): 33–41.

Sandor, Jonathan A. "Biogeochemical Studies of a Native American Runoff Agroecosystem." *Geoarchaeology* 22 (2007): 359–386.

Tomky, Naomi. "Mexico's Famous Floating Gardens Return to Their Agricultural Roots." *Smithsonian Magazine,* January 31, 2017. https://www.smithsonianmag.com/travel/mexicos-floating-gardens-return-their-agricultural-roots-180961899/.

Veys, Fanny Wonu. *Mana Māori: The Power of New Zealand's First Inhabitants.* Leiden: Leiden University Press, 2010.

Chapter 9. Soil Heroes

Binam, Joachim N., Frank Place, Antoine Kalinganire, Sigue Hamade, Moussa Boureima, Abasse Tougiani, Joseph Dakouo, et al. "Effects of Farmer Managed Natural Regeneration on Livelihoods in Semi-Arid West Africa." *Environmental Economics and Policy Studies* 17 (2015): 543–575.

Center for Human Rights and Global Justice. *Every Thirty Minutes: Farmer Suicides, Human Rights, and the Agrarian Crisis in India.* New York: NYU School of Law, 2011.

Damiano, Luis, and Jarad Niemi. *Quantification of the Impact of Prairie*

Strips on Grain Yield at the Neal Smith National Wildlife Refuge. Ames: Iowa State University Department of Statistics, 2020.

Elleby, Christian, Ignacio Pérez Domínguez, Marcel Adenauer, and Giampiero Genovese. "Impacts of the COVID-19 Pandemic on the Global Agricultural Markets." *Environmental and Resource Economics* 76 (2020): 1067–1079.

Estabrook, Barry. "Meet Allan Savory, The Pioneer of Regenerative Agriculture." *Successful Farming,* March 8, 2018. https://www.agriculture.com /livestock/cattle/meet-allan-savory-the-pioneer-of-regenerative -agriculture.

Hitch, Gregory. "Lessons from Coon Valley: The Importance of Collaboration in Watershed Management." Aldo Leopold Foundation, July 23, 2015. https://www.aldoleopold.org/post/lessons-from-coon-valley-the -importance-of-collaboration-in-watershed-management/.

Kassam, A., T. Friedrich, and R. Derpsch. "Global Spread of Conservation Agriculture." *International Journal of Environmental Studies* 76 (2019): 29–51.

Kim, Nakian, María C. Zabaloy, Kaiyu Guan, and María B. Villamil. "Do Cover Crops Benefit Soil Microbiome? A Meta-Analysis of Current Research." *Soil Biology and Biochemistry* 241 (2020): 107701.

Phillips, Ronald E., Grant W. Thomas, Robert L. Blevins, Wilbur W. Frye, and Shirley H. Phillips. "No-Tillage Agriculture." *Science* 208 (1980): 1108–1113.

Plourde, James D., Bryan C. Pijanowski, and Burak K. Pekin. "Evidence for Increased Monoculture Cropping in the Central United States." *Agriculture, Ecosystems, and Environment* 165 (2013): 50–59.

Seufert, Verena, Navin Ramankutty, and Jonathan A. Foley. "Comparing the Yields of Organic and Conventional Agriculture." *Nature* 485 (2012): 229–234.

Stanley, Paige L., Jason E. Rowntree, David K. Beede, Marcia S. DeLonge, and Michael W. Hamm. "Impacts of Soil Carbon Sequestration on Life Cycle Greenhouse Gas Emissions in Midwestern USA Beef Finishing Systems." *Agricultural Systems* 162 (2018): 249–258.

Stinson, Liz. "World's Largest Rooftop Urban Farm to Open in Paris Next

Year." Curbed, August 15, 2019. https://www.curbed.com/2019/8/15
/20806540/paris-rooftop-urban-farm-opening.

Willer, Helga. "Organic Market Worldwide: Observed Trends in the Last
Few Years." Bio Eco Actual, October 3, 2020. https://www.bioecoactual
.com/en/2020/03/10/organic-market-worldwide-observed-trends-in
-the-last-few-years/.

Chapter 10. A World With Soil

Carson, Rachel. *Silent Spring.* Boston: Houghton Mifflin, 1962.

Chambers, Adam, Rattan Lal, and Keith Paustian. "Soil Carbon Sequestration Potential of US Croplands and Grasslands: Implementing the
4 per Thousand Initiative." *Journal Soil Water Conservation* 71 (2016):
68A–74A.

4 Per 1000 Initiative (website). 4 per 1000. https://www.4p1000.org.

McGreavy, Bridie, and Laura Lindenfeld. "Entertaining Our Way to Engagement? Climate Change Films and Sustainable Development Values."
International Journal of Sustainable Development 17 (2014): 123–136.

Minasny, Budiman, Brendan P. Malone, Alex B. McBratney, Denis A. An-
gers, Dominique Arrouays, Adam Chambers, Vincent Chaplot, et al.
"Soil Carbon 4 per Mille." *Geoderma* 292 (2017): 59–86.

National Research Council. *Climate Intervention: Carbon Dioxide Removal
and Reliable Sequestration.* Washington, D.C.: National Academies Press,
2015.

Schlesinger, William H., and Ronald Amundson. "Managing for Soil Carbon Sequestration: Let's Get Realistic." *Global Change Biology* 25 (2019):
386–389.

*The State and Future of U.S. Soils: Framework for a Federal Strategic Plan for
Soil Science.* Subcommittee on Ecological Systems, Committee on Environment, Natural Resources, and Sustainability of the NSTC (December 2016). https://obamawhitehouse.archives.gov/sites/default/files
/microsites/ostp/ssiwg_framework_december_2016.pdf.

Sunstein, Cass R. *How Change Happens.* Cambridge, Mass.: MIT Press,
2019.

Von Burg, Ron. "Decades Away or the Day After Tomorrow?: Rhetoric, Film, and the Global Warming Debate, Critical Studies in Media." *Critical Studies in Media Communication* 29 (2012): 7–26.

Winsten, Jay A. "Promoting Designated Drivers: The Harvard Alcohol Project." *American Journal of Preventative Medicine* 10 (1994): 11–14.

Index

Figures are indicated by "f" following page numbers. The inserted plates are indicated by plate number (e.g., pl. 1 indicates plate 1).